GLOBAL DIGITAL HUMAN RIGHTS

in the implementation of the Global Digital Compact

The Toolkit for Human-Centered GovTech, AI and Global Governance

by Maksim Burianov

Master of Laws • SDG Ambassador • Global Shaper (WEF)

Burianov M. Global Digital Human Rights in the Implementation of the Global Digital Compact: The Toolkit for Human-Centered GovTech, AI and Global Governance. United Humans Press, 2025. 402 p.

Synopsis

This groundbreaking monograph advances the United Nations' Global Digital Compact (GDC) by translating its vision into a techno-legal standard for global digital human rights and children's rights—so that digitalization is genuinely human-centered. It interrogates the widening gap between exponential technologies and slow-moving law, focusing on AI safety, digital militarization, surveillance capitalism, inequality, and the specific vulnerabilities of children and youth. Bridging public, private, technical, legal and civil-society perspectives, the book sets out global, regional and local roadmaps for innovation ecosystems built on dignity rather than data extraction. As a first actionable step, it proposes a UN Declaration and Convention on Global Digital Human Rights to extend—and operationalize—the Global Digital Compact adopted by the UN General Assembly on 22 September 2024 as Annex I to the Pact for the Future.

Use cases

Policymakers • GovTech & AI architects • Legal scholars • Human-rights approaches and advocates • ESG strategists • Web 3.0 and network-state builders • Graduate courses in Constitutional Law, Human Rights, Digital Law, AI Ethics, Sustainable-Development Governance.

Keyword: Global Digital Compact; digital human rights; children's rights; AI governance; Web 3.0; metaverse; network state; GovTech; sustainable development; Fourth Industrial Revolution.

ISBN 979-8-9990219-0-8

Third print edition, 2025 · United Humans Press

Content

Digital Rights Foreword (by Dr. Vinton G. Cerf, the father of the Internet)

Foreword to Maksim Burianov

Global Digital Human Rights in the Implementation of the Global Digital Compact

This book wrestles with a modern conundrum: how do rights emerge in the digital realm and how do they relate to the historical rights conceived to apply in the physical world? We should first agree, for purposes of discussion, that the digital world is no less real than the palpable one we evolved in. It is part of our world, increasingly so. We live in the digital environment, encountering it concurrently with the physical world. Increasingly, actions taken in the digital realm have real world consequences. As AI, autonomous vehicles and robots of all kinds become commonplace, the digital and physical worlds will merge in a blend of both.

Burianov cites Luciano Floridi's concept of *information organisms* that concurrently inhabit the physical and digital worlds. This cohabitation has a direct effect on our perceptions and experiences. Digital encounters are no less real than physical ones and we seem to be on our way to a merging of the two. As we explore the question of rights, we must take into account this modern development. Computers are changing everything.

The notion of rights is an artifact of society. Without society, the concept of rights seems vacuous. Rights emerge as a consequence of societal norms and agreements. They are distinguished from privileges. Rights apply, in theory, to everyone, while privilege is granted to selected members of society. Rousseau's *Social Contract* teaches us that as we form societies, we may accept enforceable limitations on our behavior in exchange for benefits such as safety, security, fair treatment and general wellbeing. The framers of the American Declaration of Independence characterized this concept briefly:

"We hold these truths to be self-evident, that all men are created equal, that they are endowed by their Creator with certain unalienable Rights, that among these are Life, Liberty and the pursuit of Happiness."

Subsequently, the American society created a constitution and adopted additional amendments (*Bill of Rights)* to further enumerate the protections and benefits intended for all members of the society. In this new digital age, Burianov catalogues efforts to construct a new expression of rights to include physical and digital components and to make this formulation compatible and coherent across both realms.

A major focus of Burianov's work, in addition to compiling a comprehensive summary of the many efforts to characterize digital rights, are the rights of children. In most of today's societies, this refers to people under the age of 18. It should be remembered, however, that in earlier times, lifetimes were shorter and what we would consider adult responsibilities today were undertaken at earlier ages. For example, Alexander the Great, at age 16, successfully undertook military leadership in his father's absence.

As is well known, many species that reproduce by live birth nurture their young for a longish time, protecting them from harm, feeding them and showing them how to survive. Even birds care for their young until they are able to fly on their own. Surviving in the modern world means learning society's norms and practices. Commonly, this is achieved through education for a period of time, in addition to learning about society's ways by observation and experience. As we experience the digital world, there is a similar need to learn how it works, how to survive and how to thrive within its framework.

My reading of this book reinforced a long-held belief that we should teach young people how to survive and thrive in the digital world. We take great pains to teach young people the rules of the road and insist that they demonstrate understanding before we give them the privilege of driving in traffic. Maybe we should consider developing an Internet Driver's Course with a test that has to be passed before access to the Internet is granted! While I doubt this could actually work, the coursework seems possible and even desirable as part of preparing young people for life in our digitally enabled world.

As much as we might wish it, we cannot absolutely guarantee protection of the young from harm. They need to learn how to respond to emergencies and how to adopt safe behaviors in a hazardous world. We can equip them with knowledge and we can also try to adopt societal norms, laws and practices aimed at protection until they are ready to make their own decisions and to fend for themselves. Giving young people increasing ability to take responsibility for their actions should be an important part of helping them to grow into responsible citizens.

The Internet, World Wide Web, computers and smart phone technologies form a considerable part of the environment in which young people grow up in the 21st century. Myriad applications give us entertainment, social connection, affordances and learning opportunities. We use the digital world, increasingly, to engage with the physical world. I think it is broadly appreciated that the affordances of the digital and online world are broad and largely beneficial. Access to information at our finger tips, sharing of what we know to benefit others, discovering similar interests and collaborating to further them, learning in real time at need (think

YouTube!), convenience of online planning and execution (think holidays, shopping, appointments), the list grows long.

The same technologies that have brought enormous benefits to people who already have access to the technology (about ⅔ of the world) and even to those who don't (think vaccines against COVID developed with the help of online information and coordination), are also available to people who do not have society's best interests in mind. The amplifying power of computers and the Internet are equally accessible to everyone, criminals included. This is often the price we pay for powerful technologies. The 250 HP car can take you on an adventurous holiday and it can be used to ram, kill and maim crowds of people. The social medium that lets you share happy photos with friends also lets you share videos of beheadings and torture.

It is easy to blame the technology for many of these harmful examples but it is important to recognize that they are largely driven by people with harmful intent. Half a century ago, Douglas Engelbart and J.C.R. Licklider promoted research into artificial intelligence and the use of computing technology to augment human capability. We are living with the results of that prescient endorsement of technology. I am a believer in that mantra, but I cannot avoid the fact that the augmentation is not confined to beneficial outcomes. These are increasingly powerful tools.

Socially responsible use of technology also demands that we find ways to constrain or inhibit harmful uses or, at least, to hold parties accountable for misuse. The Internet makes this harder in some ways. It was designed to interconnect networks that, in some cases, span the globe of Earth. The design is agnostic with regard to national boundaries, by intent. The consequence is that victims of digital abuse may be in one legal jurisdiction and perpetrators in another. We still live in a Westphalian world in which geopolitical boundaries count and national laws are not guaranteed to be compatible across those boundaries.

In spite of the variations, there are attempts to create mechanisms for collaborative law enforcement (think Interpol, Budapest Convention on Cybercrime). These may be multi-lateral or bi-lateral agreements. A critical component of accountability is the ability to identify parties engaged in harmful and/or illegal behavior. This gets to the crux of a Gordian knot: identity, privacy, anonymity. To hold parties accountable, we must know who they are. We value anonymous use of the Internet but wrestle with the accountability issue. Under some conditions, we have to pierce the veil of anonymity to hold parties accountable.

This recognition has led to debates about the use of cryptography to protect privacy vs the ability of law enforcement to break codes to expose abuse and bring lawbreakers to justice. Proposals for "back doors" to cryptographic systems surface from time to time. As a computer scientist, I have argued against such designs because they invariably leak and lead to unintended violation of legitimate privacy. Some of the worst offenses are often used to argue for such back doors. Child pornography is among the worst I can think of. I have noted, however, that infiltration and other practices that gain access to communication have led to rounding up of offending parties without introducing back door cryptosystems. Brute force and other cryptoanalytic tools have also had some successes.

Ironically, strong authentication is becoming an important element for protection of safety. Making it hard for someone to pretend to be you in the online world is good practice. Multi-factor authentication (think username, password, hardware token) is one way to reduce risk. Digital signatures help to establish provenance and to verify integrity of digital objects and transactions. I believe that authentication tools and technologies will become critical parts of the digital ecosystem we want. Two examples are found in India (think Aadhar) and Estonia (think citizen's digital ID). Certificate Authorities (signing identify documents) and use of public key cryptography can play a role, although both are at risk. Certificate Authorities can be and have been compromised to produce false identities. Current public key cryptographic systems are at risk to quantum computing (think Shor's algorithm) but, by good planning, there are post-quantum cryptographic algorithms resistant to this attack already defined and in use.

Burianov explores the potential of the recently adopted Global Digital Compact. While it is only a framework that needs substantial populating with specific international agreements, it, along with the earlier Budapest Convention on Cybercrime and the recent Cybercrime Convention, may be a scaffold on which better citizen protections can be erected. A significant element of Burianov's concerns lies with protection of children from harm. Some of the concern surrounds misinformation and disinformation in the digital realm. Such problems have been around since the dawn of humanity and, in my opinion, no technology will rid us of them. However, we can teach young and old to think more critically about what they see and hear in the virtual and physical worlds. We should naturally ask questions about provenance, supporting and refuting information, healthy skepticism and other practices that teach us to scrutinize assertions before accepting them at face value.

Every new technology brings its own challenges and the online world with its powerful computing engines is no different. What is important is that explorations like the one in this book help us to adapt to these challenges and adopt practices that emphasize their beneficial properties while reducing the risks they pose.

Vinton G. Cerf

Woodhurst

February 2025

Digital Rights Foreword (by A. Kalinina, Co-Founder and CEO, reState)

Global Digital Human Rights in the Implementation of the Global Digital Compact Foreword
by

Anastasia Kalinina, Co-Founder and CEO, reState Foundation, reimagining the future of governance and global collaboration

In the beginning, there was not code.
In the beginning, there was intention.

The sacred intention of life – to become more whole, more conscious, more true to itself through evolution, expression, and love. Technology, when born from this frequency, becomes not a tool of dominance, but a bridge to liberation.

This book is not a manifesto. It is a map.
Not a doctrine. But a declaration of possibility.
A compass for those of us – quietly or boldly – already navigating the new epoch of governance, identity, and human becoming.

Maksim Burianov does something rare: he dares to reimagine the architecture of global civilization through the soul of the child and the sovereignty of the human being. At the heart of his vision lies a simple but radical proposition: that digital transformation must be human transformation. That global compacts, data protocols, and AI governance only matter if they serve life, dignity, and joy.

We stand at a civilizational threshold. The 21st century is not asking for reform – it is asking for rebirth. And like all births, it is messy, luminous, and entirely dependent on who we believe ourselves to be.

Are we algorithmic units in a planetary machine?

Or are we creators, co-designers of planetary harmony, attuned to the subtle intelligence of love?

This book invites us to choose.

It invites us to leave behind the outdated saddle of nation-states and legal inertia, and instead board the interstellar craft of distributed, human-centered governance. It names the distortions – digital inequality, data colonialism, ethical confusion – and offers something better: a Declaration of Global Digital Human Rights not as legal formality, but as a generative source code for civilization itself.

It calls to the young – to the "inforgs," the digital natives who will code, dream, and legislate the future – not as technocrats, but as sacred stewards of what it means to be human in an age of superintelligence. It reminds us: our technologies will be no wiser than the consciousness that births them.

This is a book for visionaries. For systems architects. For protectors of childhood and champions of the commons. For those daring to code love into law.

May you read this as a compass, not a conclusion. May it provoke you to ask:

What would a world look like where our rights begin with reverence?
Where every child's access to the digital realm is not a transaction – but a sacred trust?
Where data is not mined, but honored?
And where our collective intelligence finally learns to serve our collective soul?

The future is not something we enter.
It is something we remember.
And then choose to build – together.

Foreword 2025. Against the Noise: The Human among Probabilistic Models

They work with data, information and noise. They monetize our cognitive fingerprints and replace the human with probabilistic strings of text. More and more, we edit someone else's generation instead of creating our own. Machines answer and choose, analyze and compile; grain by grain they take our skills and our time. This is social mining—a new phase of surveillance-platform techno-capitalism in which the proletariat of code, text, likes and clicks extracts value for platforms that speak our language. Generative AI does not think; it produces signs that we mistake for meaning. We see the façade of words and supply the content ourselves. We reach for large language models to do more and think faster—only to find ourselves unlearning how to think.

In an age of noise and the façade of meaning, we face two paths. The first is to learn to manoeuvre in the data world of AI: to use tools as tools rather than become their extensions; to recognize where algorithms nudge our choices and where "our" voice is only a mirror of platform bias. The second is to build systems that are human-centered by design—systems that return freedom, joy and the right to one's own voice, and in which digital practices amplify, rather than displace, our capacity to think and create.

These challenges have many names: "*surveillance capitalism*" (Shoshana Zuboff), "*technofeudalism*" (Yanis Varoufakis), "*platform capitalism*" (Nick Srnicek). Yuval Noah Harari writes about the ideology of *dataism* and about humans becoming "*hackable animals*", vulnerable to data-driven manipulation. In the language of international organizations: close the governance gap in digital technologies and AI, and uphold the principle that the same rights people have offline must be protected online. But this book is not only a diagnosis. It sets out an architecture of action: a *Declaration and Convention on Global Digital Human Rights;* a *Digital Rights by Design standard;* new institutions—digital ombudspersons, independent AI auditors, transparency registries for data and algorithms—and effective mechanisms for complaint and remedy. The aim is simple: for technology, by design, to expand freedom, choice, joy and opportunity for everyone—especially children—rather than entrenching inequality and control.

DISCLAIMER

This book solely represents the author's opinions and does not reflect the official positions of any organizations or institutions. All ideas, analyses, and conclusions presented in the book are based on the author's personal views and are intended to stimulate discussion and deepen the understanding of the topic of global digital human rights. The author is not responsible for any interpretations or consequences resulting from the use of the information provided.

For Collaboration: globalmaxlaw@gmail.com

DIGITAL FORK IN HISTORY

We stand on the threshold of an era in which science-driven progress is reshaping the very foundations of our lives. Tectonic shifts in the economy, the fabric of social relations and the scale of global challenges have become part of daily reality. Amid the explosive growth of digital platforms, one question dominates: how do we preserve—and fully realize—human dignity and opportunity in a world where the rules are written in code?

A proposal for a Declaration of Global Digital Human Rights was first introduced in 2019-2020 as a new social contract for the digital age. The concept gained traction at the World Economic Forum and later informed the Global Digital Compact adopted by the United Nations in 2024. While the GDC established a baseline, it addresses barely half the actions needed for a truly human-centred digital transformation. Its third objective— *"create an inclusive, open, safe and secure digital environment where human rights are respected, protected and promoted"*—calls for end-to-end safeguards across the full life-cycle of emerging technologies and endorses the UN Guiding Principles on Business and Human Rights for the private sector.

Yet the community still lacks (i) a comprehensive catalogue of digital rights, (ii) concrete enforcement mechanisms and (iii) a deeper grasp of how fast-moving technologies will shape human self-realization. The GDC is a starting point, not a finish line. Actionable steps—outlined in this book—remain essential.

Filling the normative gap: from the mid-twentieth century to the AI age

The *International Bill of Human Rights*—the 1948 Universal Declaration and the two 1966 Covenants—codified universal values during the second and third industrial revolutions but never captured their digital dimension. Recent UN outputs underscore that omission:

- **A/HRC/50/55 (2022)** OHCHR report on deliberate Internet shutdowns[1];
- **A/HRC/53/42 (2023)** OHCHR study on human-rights considerations in technical standard-setting[2];
- **A/RES/78/216 (2023)** GA resolution calling for clear guidance on digital technologies[3].

Since 2012 the UN has begun to close the gap through a sequenced body of resolutions:

Year	UN Instrument	Key contribution
2012	**A/HRC/RES/20/8[4]**	Launches the *"offline = online"* principle for HR.
2013	**A/RES/68/167[5]**	Condemns mass surveillance; affirms digital privacy.
2020	**A/RES/75/176[6]**	Introduces mandatory human-rights impact assessments.

[1]Internet shutdowns: trends, causes, legal implications and impacts on a range of human rights Report of the Office of the United Nations High Commissioner for Human Rights (13 May 2022).

[2]Human rights and technical standard-setting processes for new and emerging digital technologies: report of the Office of the United Nations High Commissioner for Human Rights (18 September 2023).

[3] Promotion and protection of human rights: human rights questions, including alternative approaches for improving the effective enjoyment of human rights and fundamental freedoms / URL: https://documents.un.org/doc/undoc/gen/n23/421/92/pdf/n2342192.pdf

[4] The promotion, protection and enjoyment of human rights on the Internet : resolution / adopted by the Human Rights Council / URL: https://digitallibrary.un.org/record/731540?v=pdf

[5] The right to privacy in the digital age / URL: https://docs.un.org/en/A/RES/68/167

[6] Promotion and protection of human rights: human rights questions, including alternative approaches for improving the effective enjoyment of human rights and fundamental freedoms / URL: https://docs.un.org/en/A/RES/75/176

2023	**A/HRC/RES/53/22**[7]	Declares some AI uses (e.g., mass facial recognition) *incompatible* with HR.
2023	**A/RES/78/213**	Umbrella text safeguarding encryption and codifying HR due-diligence across tech lifecycles.
2024	**A/RES/78/265**[8]	Seizing the opportunities of safe, secure and trustworthy AI systems for sustainable development. Adopted without a vote (A/78/PV.63). Emphasizes that human rights and fundamental freedoms must be respected, protected and promoted throughout the AI lifecycle; calls to refrain from or cease use of AI systems incompatible with international human rights law or creating undue risks, especially for vulnerable groups; reaffirms that rights offline must also be protected online.
2024	**A/RES/79/239**[9]	Affirms that international law — including the UN Charter, international humanitarian law, and international human rights law — applies throughout the entire life cycle of AI capabilities, as well as to AI-enabled systems in the military domain.
2025	**A/RES/79/325**[10]	Ensured the establishment and functioning of the Independent International Scientific Panel on

[7] Promotion and protection of human rights in the context of digital technologies / URL: https://documents.un.org/doc/undoc/gen/n23/422/28/pdf/n2342228.pdf
[8] https://docs.un.org/A/78/L.49
[9] https://documents.un.org/doc/undoc/gen/n24/426/98/pdf/n2442698.pdf
[10] https://docs.un.org/A/79/L.118

		Artificial Intelligence and the Global Dialogue on Artificial Intelligence Governance, including the respect, protection, and promotion of human rights in the field of artificial intelligence.

Together, these resolutions and technical standards supply the missing architecture the *International Bill* never foresaw. They turn abstract rights into programmable parameters—from bandwidth affordability to algorithmic transparency—yet they still stop short of a consolidated charter with teeth.

This book argues that digital human rights must become the new performance indicators of the digital state—benchmarks of efficiency, accountability, transparency, decentralization and true human-centeredness. Embedding those indicators in gov-tech, civ-tech, DAOs and other Web 3.0 architectures will make public action proactive and inclusive, leaving no one behind and rendering every individual's participation genuinely meaningful. Books by Global Shaper (WEF) and SDG Ambassador Maksim Burianov "Digital Human Rights in the Context of Global Processes"[11]and "Digital Rights of the Child"[12]explores the future of digital human and child rights in the era of rapid technological progress.

The pages that follow offer a roadmap for translating today's scattered resolutions into an enforceable, twenty-first-century catalogue of digital rights—and a techno-legal toolkit robust enough to protect them.

Testing the idea

Enshrining digital human rights can accelerate the transition to a human-oriented digital state. The e-government index demonstrates a correlation between countries that have enshrined digital human rights to access the Internet in their constitutions with an increase in the level of e-government (Estonia, considered one of the leaders in the Govtech

[11]https://book.ru/book/943404
[12]https://book.ru/book/952736

sphere, ranks 8th in the index; in 2000, the Estonian parliament passed legislation and enshrined Internet access)[13].

In support of the adoption of the Declaration of Global Digital Human Rights, the concept was featured in the *World Economic Forum*[14] *Global Agenda* and accompanied by a public petition. Over the past seven years, it has been tested and debated in more than 100 scientific and international conferences, including high-level UN youth events.

The framework also informed the legal component of the Public Administration Ecosystem Reference Architecture (PAERA)[15] within the ITU-led GovStack initiative, highlighting the role of human rights in digital transformation and governance.

Most recently, the draft Declaration was presented during the 2025 Global Digital Compact Townhall at UN Headquarters and in the human rights track of WSIS+20.

Statement of the Problem. Scientific research on digital rights increasingly highlights the need for a new techno-legal standard—one capable of aligning legal norms with the rapid evolution of digital technologies. The current legislative lag not only limits the use of emerging opportunities but also reinforces inequalities in access to digital resources, deepening socio-economic divides and constraining individual self-realization.

This study focuses on the development of practical tools, concepts, and action plans to help societies adapt to the challenges of the digital era. Central to this vision is the protection of human rights, particularly the rights of children—in the context of global digitalization. Advancing towards the Sustainable Development Goals by 2030 will require consistent progress on the adoption of a Declaration and, ultimately, a Convention on Global Digital Human Rights.

---

[13]UN Study: E-Government 2022 / https://desapublications.un.org/sites/default/files/publications/2023-01/UN%20E-Government%20Survey%202022%20-%20Russian%20Web%20Version.pdf.

[14]https://www.weforum.org/agenda/2020/08/here-s-why-we-need-a-declaration-of-global-digital-human-rights/

[15] https://govstack.gitbook.io/paera-doc

Box 1. Key dilemmas of the digital transformation and proposed solutions

The digital transformation poses fundamental dilemmas for the realization of human rights and the protection of children online. Each of these dilemmas requires clear responses that balance innovation with human dignity and global equity:

1. **Natural vs. Artificial Intelligence** – *Dilemma:* How to coexist with rapidly advancing AI. *Solution:* Strengthen systemic and critical thinking in people, while requiring AI systems to reflect humanistic values grounded in human rights.
2. **Robot Rights vs. Human Rights** – *Dilemma:* Whether AI should be granted rights. *Solution:* Affirm that rights belong exclusively to human beings and prevent AI—including AGI and ASI—from being given independent legal status.
3. **Digital Security vs. Digital Freedom** – *Dilemma:* How to protect children without restricting access. *Solution:* Ensure children's access to digital technologies for learning and development, while embedding child-safety mechanisms at the platform level.
4. **Self-Regulation vs. Public Regulation of BigTech** – *Dilemma:* Whether corporations can govern themselves. *Solution:* Establish human-centred regulatory frameworks that balance private sector innovation with the broader interests of the global community.
5. **Digital Citizenship vs. National Citizenship** – *Dilemma:* Identity in a hyperconnected world. *Solution:* Promote global networked citizenship rooted in dialogue and cooperation to address shared challenges beyond national borders.

6. **Digital Identity vs. Human Individuality** – *Dilemma:* The risk of reducing people to data profiles. *Solution:* Safeguard the uniqueness and creativity of the individual while integrating digital identities into social and legal systems.
7. **Centralization vs. Decentralization of Governance** – *Dilemma:* Competing models of control. *Solution:* Develop hybrid governance systems that combine the efficiency of centralization with the resilience of decentralization, anchored in digital identity, human rights, and sustainable development.
8. **Classical vs. "Clip" Thinking** – *Dilemma:* How to reconcile deep analysis with fragmented digital attention. *Solution:* Combine systemic and critical reasoning with creative thinking for human–AI co-creation.
9. **Ethics vs. Law** – *Dilemma:* Whether ethical guidelines are enough. *Solution:* Create a global techno-legal standard that embeds human rights and child rights into binding legal frameworks while drawing on the highest ethical principles.

These challenges demand a comprehensive, human-centric approach to managing digital technologies and supporting sustainable development. A key step is to link global digital human rights with the idea of a **Global Digital Commons**—treating access to computing power, platforms, knowledge, data, and essential services as public goods. Digital infrastructure should not be seen as private or fragmented assets, but as resources collectively governed for equity and cross-border cooperation.

Digital Rights by Design provides a pathway to ensure that future Artificial General Intelligence (AGI) and Artificial Superintelligence (ASI) serve the interests of humanity as a whole. Embedding human rights principles directly into the architecture of these technologies can safeguard privacy and sovereignty, prevent misuse, and promote solutions to shared global challenges. This approach places dignity and inclusivity at the centre of innovation. It aims to create technologies that enable equitable growth, support collective well-being, and uphold ethical standards. The concluding sections of this book outline proposals for managing the transition to singularity, with frameworks that ensure human-centric development in the emerging AI era.

INTRODUCTION: BRIDGING THE DIGITAL VOID FROM DECLARATION TO TECHNICAL STANDARD DIGITALRIGHTSBYDESIGN

As work progresses on the draft **Declaration of Global Digital Human Rights**, it is important to outline the next stages in integrating these rights into the hierarchy of international norms. Building on existing legal frameworks, this incremental approach will help embed digital rights within the global legal order.

The Declaration should be considered the foundation—a first step toward global recognition of digital rights. From there, complementary mechanisms are needed: international agreements, regional treaties, and binding technical standards to ensure practical enforcement.

The subsequent stage is to position digital rights within the broader hierarchy of international law, identifying entry points in current frameworks and opportunities for innovation. The table below summarizes the levels of international norms and indicates where digital rights can and should be incorporated.

Box 2. Hierarchy of Norms in International Law

Level	Summary	Examples	Incorporating Digital Rights
1. Jus cogens (Imperative Norms)	Overriding norms from which states cannot deviate; "common interests of humanity."	Prohibition of genocide, slavery, aggression.	Digital rights could theoretically be included as a fundamental *right to human dignity in the digital age*, but this would require sustained state practice and opinio juris.
2. UN Principles	The UN Charter takes precedence over all treaties (Article 103).	Non-intervention, sovereign equality.	The digital realm can be reflected in UN resolutions like A/RES/78/213 → the next step would be an amendment to the Charter to include digital neutrality, ensuring that critical infrastructure and digital infrastructure cannot be disrupted by states.

3. Erga Omnes Obligations	Duties towards the international community as a whole; arise from jus cogens and key treaties.	Right to self-determination, basic human rights.	"Access to technology as a right" could evolve into *erga omnes* if states begin referencing it in disputes and courts.
4. Universal Human Rights Treaties	Legally binding pacts; monitored by the CCPR / CESCR.	ICCPR, ICESCR.	A new **Declaration** of **Global Digital Human Rights** (guaranteeing fundamental rights, new digital rights, and rights embedded in code), followed by a **Convention on Global Digital Rights** would help realize these rights.
5. Other Universal Treaties	Thematic conventions approved by the UN General Assembly.	Convention on the Rights of the Child, Convention 108+ (CoE ↔ Global Access).	We could initiate a **Global Convention on the Right to Universal and Meaningful Connectivity**, to accelerate overcoming the digital divide and ensure access to Industry 3.0 for all.

6. Regional Treaties & Courts	Take precedence over national law in participating countries.	European Convention on Human Rights, African Charter, ACHR.	Digital articles should be integrated into protocols to the ECHR / African Charter (this is already under discussion).
7. Customary International Law + General Principles	Practice + opinio juris → norm; "gap-filler" in the absence of treaties.	Diplomatic rules, immunities.	The ongoing reference to "offline = online" (A/HRC/RES/20/8) is gradually forming customary law, as states increasingly recognize digital rights.
8. Soft-law and Technical Standards	Norms of practical application; informally guide behavior.	Resolutions from OHCHR/GA, guiding principles, codes of conduct.	A detailed **code of implementation for digital rights** should be developed as part of international soft-law, complementing the formal treaty processes.
8.1 Intergovernmental Standards	ITU-T, ISO/IEC, WHO-ITU: specifications, recommendations, indicators.	**PP-18 Res 139 (UMC)**, **ITU-T Y.2770**, **H.870**, **FG-AI4A**.	These should be integrated into new ITU-T series (e.g., Y.Digital-Rights) and ISO/IEC standards on **Human-Centred AI**.

The development of digital human rights must follow a structured path, starting with the **Declaration of Global Digital Human Rights** as the central framework, complemented by UN resolutions, regional conventions, binding technical standards, and soft-law initiatives. The above table outlines the necessary levels of integration, from imperative norms (jus cogens) down to technical standards in industry platforms and multi-stakeholder organizations.

By working through these steps systematically, we can create an integrated, robust, and enforceable system of digital human rights that addresses the challenges of the digital age while ensuring that no one is left behind.

Digital Rights by Design. Envisioned as a cross-sector **meta-standard,** *Digital Rights by Design* would sit "one layer above" functional specifications, embedding human-rights requirements into every stage of the technology lifecycle. Its architecture could mirror the ISO/IEC high-level structure (HLS) for management systems and contain five normative clauses.

At a minimum, this may include: **(1)** specific requirements for system architecture; **(2)** human-rights compliance checks and due diligence for companies and A-round and later-stage startups; and **(3)** product-certification mechanisms embedded in state digital platforms and explicitly oriented toward the realization of digital human rights rather than other objectives. Consistent with UN practice, two cross-cutting safeguards should apply across all three areas: independent oversight and accessible remedy (including, where appropriate, the use of distributed-ledger technologies at the national level to register claims, compensate harm and restore violated rights), and transparency through public reporting on compliance and incidents.

Annexes would offer sector-specific profiles—cloud services, edge devices, public-sector AI — plus an implementation toolkit harmonized with IETF/W3C best practices (e.g., RFC 8890, W3C Ethical Web). Certification bodies could leverage existing ISO/IEC 17065 frameworks, enabling regulators and procurers to mandate *Digital Rights by Design* compliance as a pre-condition for market access or public tenders, thereby operationalizing the human-centred goals of the Global Digital Compact.

With regard to AI, the technocratic approach based on granting "freedom to the machine," promoted in some corporations, shifts the focus from the human to the algorithm, turning the company into an autonomous organization where rights and freedoms yield to "optimality." Such a trajectory inevitably leads to the risk of digital dictatorship and underscores why algorithmic freedom must remain secondary to human freedom. This is precisely why embedding the principle of *by design* across the entire lifecycle of technologies—products, services, and platforms—is decisive for the realization of digital human rights.

Having outlined the hierarchy of international norms and the path towards embedding digital rights into law, it is time to step back and reflect more broadly. Legal frameworks are necessary, but they are not the whole story. To truly understand what is at stake, we must place these norms in the wider context of culture, technology, and the human experience of living through a digital revolution.

PREFACE: IN THE BEGINNING THERE WAS CODE

In many parts of our world this book would be called differently: in France it is "Droits de la Petite Poucette"[16](The rights of the little girl with a thumb), in the UK "Rights of the sovereign individual"[17], in Italy "Diritti degli information"[18](Rights of Informers), in Israel " זכויות האדם של המחר"[19](Human Rights of Tomorrow), and in America "Rights in code"[20]. In World we call it "**Global Digital Human Rights in the implementation of the Global Digital Compact**".

This is not just a book—it is an invitation to reflect on the profound shifts of our time. It speaks to everyone: ChatGPT enthusiasts, admirers of metamodern philosophy, fans of *Westworld*, digital natives, and, of course, experts in international organizations, digital lawyers, nomads, and decision-makers in major tech companies. This book serves as both a continuation and a reimagining of previous discussions on individual rights in the context of the Fourth Industrial Revolution and the mounting shadows of global challenges.

In this new work, we journey into a near future shaped by generative artificial intelligence, the decentralized architectures of Web 3.0, the augmented realities of metaverses, and the cognitive frontiers of neural interfaces. Like any compelling narrative, there is a villain, though it may not be the one you expect. This time, it is not a rogue artificial intelligence with malevolent plans.

The true villain of our age is the growing imbalance between the *rapid acceleration of technological* and social transformation and the *lagging evolution of our legal, ethical, and regulatory frameworks*. This villain is far more insidious, threatening not through overt hostility but

---

[16]Serres, M. La petite poucette. Éditions Le Pommier. 2012. – 80 p.

[17]Davidson JD, Rees-Mogg W. The Sovereign Individual: Mastering the Transition to the Information Age. Simon & Schuster. 1999. – 448 p.

[18]Floridi, L. Philosophy and Computing: An introduction. Routledge. 1999. – 257 p.

[19]Harari, YN Homo Deus: A Brief History of Tomorrow. Harvill Secker. 2016. 448 p.

[20]Lessig, L. Code and Other Laws of Cyberspace. New York. 1999. – 424 p.

through our inability to adapt to new realities. It is not a robot stealing our jobs that we should fear, but our collective incapacity to grasp and respond to the implications of exponential technologies—technologies made increasingly accessible through innovations like OpenAI, Worldcoin, Neuralink, Google Quantum AI, virtual reality education platforms, and advanced language models.

The digital world is simultaneously a realm of boundless potential and an uncharted, perilous territory. Most troubling of all, human and child rights in the digital age remain undefined and unprotected. This gap denies them equitable access to the benefits of the digital economy while exposing them to the growing dangers of digitalization: ***militarization, inequality, surveillance, and malicious AI.*** As adults, we struggle to address these issues, but it is the next generation who will bear the full weight of our decisions—or our inaction. They are the pioneers of this new frontier, confronting challenges we can scarcely imagine.

One thing is certain: the linear thinking of the past is no match for the exponential technologies of the future. We stand at a crossroads, but the path forward is not a simple line toward a digitally transformed world where poverty is reduced, inequality is overcome, and the environment is improved. Achieving such outcomes requires more than selecting the "right" direction. It demands that we fundamentally rethink who we are, what we owe to our children, and how we shape the digital landscapes they will inherit.

"**High tech, low life**"—a cyberpunk slogan born from the gritty depths of 1980s science fiction—once captured the awe and dread of the "countless" possibilities brought by the Third Industrial Revolution. It served as both a warning and an anthem, foretelling a future where towering technological marvels overshadowed the human condition.

But the future moves forward, and so do its slogans. As the 21st century unfolds, a new mantra defines our age: **"More tech, less rights".**

With a greedy gleam in their eyes, our cybernetic capitalists proclaim:

- "More smart homes, less room to roam."
- "More digital footprints, less mindful imprints."
- "More merging with the virtual, less meaning in the actual."

What began as a cautionary tale of "High tech, low life" has evolved into a stark reality—a paradoxical world where technological progress accelerates, yet human rights and freedoms erode in its shadow. Our task is not only to decode this narrative but to rewrite it, ensuring that the future of "more tech" also means more rights, more dignity, and more humanity.

Corporate digital ambitions spare no one, not even our youngest generation: **"More social screens for your kids, less biometric means that forbid."** Perhaps at this very moment, your daughter or little sister is gleefully snapping selfies on Snapchat[21], blissfully unaware that the app is storing and analyzing her facial recognition data without consent, leaving her personal information exposed to risks she cannot yet comprehend[22]. Or consider your son, immersed in the vibrant universe of Fortnite—not because games are inherently harmful, as they often foster creativity and skill development—but because of their darker commercial underpinnings.

Take Epic Games, for instance, whose gaming platforms have been accused of violating the privacy of millions of children through manipulative "dark patterns"—design tricks that nudge young users into unintended in-game purchases[23]. These actions, cloaked under the guise of "long-standing industry standards," are little more than a corporate strategy to exploit younger users, the very demographic that fuels the growth of these platforms.

While many in the younger generation explore the vast expanses of the digital world with innocent curiosity, they unknowingly surrender their rights to corporations thriving in the age of surveillance capitalism. Yet, behind the glittering façade of technological progress lies a quieter, often overlooked reality: the other side of the digital divide.

In this hidden world, over 2.5 billion people remain untouched by the terms we take for granted—"Internet," "digital rights," "blockchain," or "social media." For this sea of faces, the digital age might as well not exist. Among them are 350 million children who have not only

---

[21]Read more:https://www.cnet.com/personal-finance/youve-only-got-one-day-left-to-claim-money-from-snapchats-35-million-privacy-settlement/(date accessed: 10.07.2023).

[22]Read more: https://www.washingtonpost.com/technology/2022/05/05/snapchat-teens-nudes-lawsuit/ (accessed: 10.07.2023).

[23]Note: "Dark patterns" such as confusing button labels, hidden user consents, and deliberately crafted UI layouts that trick users into making online purchases are nearly ubiquitous in the industry. Read more:https://www.nytimes.com/2022/12/19/business/ftc-epic-games-settlement.html(date accessed: 10.07.2023).

been left behind by technological progress but have never even set foot online. For them, the vast windows to global communication, entertainment, and education remain shuttered, blocked by the remnants of the **Second Industrial Revolution**[24].

We have sprinted forward in a space marathon of technological development, yet our **legal and ethical compass** remains woefully stuck, unable to guide us toward fairness and equity. The challenge before us is monumental: to bridge not just the digital divide but also the moral and regulatory chasms that threaten to define our future.

With our eyes fixed on a bright future, we find ourselves frozen in place, gradually ceding ground to the spotlights of facial recognition cameras in smart cities—cities that seem to remember everything yet forget the most fundamental rights. How strange and ironic that our era of high technology suffers from short-term memory, like a goldfish in an aquarium—or, perhaps even more ironically, a modern artificial intelligence like ChatGPT 4.

Civilization's memory, one would hope, should be enduring. Yet it seems to have forgotten the lessons of the 20th century: the Covenants and Declarations, the uprisings and barricades, the passionate movements for freedom, equality, and fraternity. Where have those fiery slogans gone? Where is that unyielding call for justice?

Where do we turn for solutions? Strangely enough, the answer lies where the problem originated. I dive into the digital depths, enter my queries into search engines, and even consult ChatGPT in search of answers. Nothing. And so, I try again, tense fingers typing into the binary void. Once more, no results. Exhausted, I sift through countless sources, write to youth councils, address the upper echelons of power—only to be met with dismissals: **"Not in our competence."**

Then, like an unstoppable tide, the pandemic arrives, bringing new worries and accelerating the next industrial revolutions—Web 3.0 becomes the buzzword of the day. Web 3.0! What is it? Meritocracy

---

[24]Note: Industrial revolutions represent major shifts in technology and society. The first (1760–1840) marked the transition from manual labor to mechanized, steam-powered labor. The second (1870–1914) involved mass production and improved communications based on electricity and oil. The third (1960–2000) saw the transition from mechanical and analog technologies to digital ones, including the internet. Finally, the fourth (21st century) represents the confluence of digital, physical, and biological technologies, changing the face of the modern world with AI, blockchain, and quantum computing.

dressed as democracy? A vague promise? And what happened to the original web? **Vinton Cerf, where are you?**

In my search for answers, I decided to write to the so-called "Father of the Internet," Vinton Cerf. To my surprise, he was curious about my initiative. Of course, my messages at Summit Zoom conferences were persistent—though always respectful. During our exchanges, I wondered: does the creator of the Internet himself pay for its use, or does he receive royalties every time we pay for access to his creation?

In our conversation, he offered a thought-provoking observation: "Many believe that digital rights and traditional human rights should be identical."

"Identical"—what a peculiar word. Even in nature, two identical snowflakes do not exist, and in the realm of blockchain, every hash is unique. How could the rights of people in the Internet age and the pre-Internet age possibly be the same? He elaborated: "The amendments to the US Constitution offer valuable guidance on adapting rights for the Internet: privacy, freedom of assembly, freedom of speech. The European Union's introduction of the 'right to be forgotten' is another innovative step. Perhaps we should even consider a new right: **freedom from harm.**"

He recommended a book, which I later referenced in footnotes 15 and 16. I remain deeply grateful to Vinton Cerf—not only for creating the Internet, but for his advice, his insights, and his thoughtful engagement with these urgent questions.

Perhaps, when Vinton Cerf envisioned the Internet, he never imagined that we would go beyond "surfing" its virtual waves. The World Wide Web has become far more than a tool for communication and knowledge—it has united an entire global community of thinking beings in the face of shared challenges.

Problems, like nothing else, have a unique ability to unite humanity. And as our problems become increasingly global, so too must our communities. Now, more than ever, we need to build bridges across divides, connecting the voices of individuals and nations to tackle the pressing challenges of our digital age.

Opportunities arise here and now. The Internet of the present moment boldly strides into the future. From where and where to? From the center – in all directions!

From a user structure with a monopoly of corporate hobbyists to collect our data, like baseball cards once were, in the hope of selling them for more in ten years. But now we don't have to wait that long. Our data is the new oil, and therefore, you and I are as valuable as oil wells. But what happens to a well when the resource is exhausted? ChatGPT reminds us that "the first thing an oil well is to close. This is done by filling it with special materials, such as cement, to prevent the remaining oil or gas from escaping." Perhaps when humans are fully "decoded," fully understandable—or, in other words, when the era of super-intelligent AI (AGI) arrives, capable of physically interacting with the world as well as we do, or even better and faster—we humans may no longer be valuable.

You probably wanted Universal Basic Income and an early extended "retirement" at 25 years old thanks to global automation, when you will be drawing and making collages from the cuts, as bequeathed by the grandfather of cyberpunk Burroughs[25]? But that won't happen anymore, because creative AI is already writing music, creating art, and effortlessly copying your unique communication style. You happily ask it to write your work emails for you and share everything with it. In the TV series "Westworld", people were fitted with cameras and tracking devices in their hats and other clothes to collect all the data to copy their identity. And we were simply given free ChatGPT, free search engines, and social networks. Give your all - optimize to zero. To solve the problem of excess "superfluous" people - we are "cemented" into cycles of metauniversal stagnation, offering any "blue pill" to your taste, stimulating the human nervous system with a brave new world.

The future we envision could be much more accessible and positive if we introduce one critical variable that we will discover by the end of our book. We are currently in a state of transition, on the threshold of a new era of the Internet - version 3.0. These future promises to be a better place for everyone.

But what exactly are these dimensions of Internet 1.0, 2.0, 3.0? Let's figure it out before we move on. First, keep in mind the formula for progression: from "Viewers" to "Users" and eventually to "Participants". In the era of Internet 1.0, we were observers of the birth of simple digital images, including the first binary pictures of cats. Web 2.0 moved us to the position of users, allowing us to like and share images of cute cats through centralized servers. But Web 3.0 changes everything. It reshapes our role from viewers and users to active participants in this interacting network. We are no longer just watching cats - we are becoming them, and our every virtual paw print is part of this digital metaverse. Together,

---

[25]D. Khaustov. Burroughs, who exploded. Beat generation, postmodernism, cyberpunk and other fragments. Moscow: Individuum, 2020. - 320 p.

we are shaping a new reality where the importance of the participant is placed above the mere viewer, and this represents our modern turn from classical centralized ideologies. However, now, this whole process resembles Schrödinger's cat paradox - we cannot yet accurately determine its true state...

However, we dissent in all directions of the decentralized Web 3.0 network—so vast, so horizontal—that all members of our planetary community could stand side by side, aligned by their true interests and personalities, guided by the desires of their genuine hearts, not by the cryptocurrency wallets of technocratic capitalist tycoons. And if we were to turn this line upside down, it would transform into a bridge stretching to the Moon, or perhaps even to Mars—a **"decentralized vertical"** built upon shared human values.

When we create conditions where every individual has an equal impact on the planet, we will no longer need to stand on the shoulders of leviathan giants. **We ourselves will become sovereign giants,** capable of reaching any goal, crossing any orbit. And we will do so without the bitterness of an unchecked ego or the corrosive ambitions that destroy the roots of our Mother Nature. Just as Oedipus, in ignorance, struck down his father, so too does ExxonMobil, with its relentless pursuit of oil, strike at the heart of our living planet. (Note: ExxonMobil has been a leading force in climate change denial, opposing regulations to curb global warming.)

In this reckless pursuit, we forget the essence of our virtues, sacrificing our human rights to metauniverse dreams that serve corporate reality rather than collective humanity.

Today, there are over 7,000 languages spoken on Earth. The Universal Declaration of Human Rights, adopted in the aftermath of World War II, has been translated into roughly 500 of them—a mere 7%. Now, it is time to translate the Declaration into yet another language: the language of code. By doing so, we can empower every person on this planet to move from being a spectator of tectonic digital transformations to becoming a creative digital author of their future.

The binary world of ones and zeros will open gates to infinite possibilities for every individual. That is why we focus on crafting a positive image of the future—one that shows how we can overcome the dichotomies born from the clash of the old and new worlds. **Robot rights vs. human rights. Code-defining law vs. code as law. Digital protection vs. digital freedom. Self-regulation vs. total regulation. National citizenship vs. digital citizenship. Individuality vs. digital**

identity. Centralization vs. decentralization. Pre-internet thinking vs. clip thinking. Ethics vs. law.

This book charts the digital steps we must take to erase invasive digital traces, narrow the lens of surveillance, and expand the horizons of cyber freedoms—from childhood onward.

In this story, girls and boys born digitally, carrying the world in the tips of their thumbs, navigate adolescence as information organisms—discovering themselves in a networked world. By adulthood, they evolve into conscious, sovereign individuals of the Web 3.0 generation. Armed with digital compasses and enshrined in global digital rights, they will unlock unprecedented opportunities to co-create a shared future. As they mature into global citizens, they will engage in techno-legal platforms of participation, co-manage planetary development, and collaborate with artificial agencies.

These new leaders will create new networked communities, cities, connections, and governance based on ***unconditional love***, not total expansion. Technology, as an extension of their ***right to develop***, will help us with this.

METHODOLOGICAL PREFACE

We live at a turning point in history. Digital networks now shape our economies, cultures, and daily lives, yet the values and laws that guide society remain rooted in a slower age. This growing gap is more than a technical problem—it is a moral and political one. The recognition of digital human rights is essential to ensure fair access to technology and the benefits it brings. Without them, the very tools designed to connect us risk deepening inequality and eroding trust.

A new global divide emerges—a division between digital elites and the "unconnected," stretching across the cybernetic axis of the world. This imbalance grows ever deeper. In this new age, it is no longer gold or land, but access to information and technology that defines one's position in society. For instance, according to UNESCO, nearly 90% of all neurotechnology patents registered with major intellectual property offices originate from just six countries.

Meanwhile, the new digital world casts a majestic shadow over the planet—a shadow of technological, informational, and economic dominance wielded by a select few over the many. This shadow is not just a consequence of innovation; it is a manifestation of systemic imbalance, where the promise of a connected world increasingly reflects the unequal concentration of power, resources, and opportunity.

To chart a path forward, this book will examine the intersections of these challenges and opportunities, advocating for a recalibration of values, laws, and governance structures to meet the demands of the digital era. Only by addressing these imbalances can we create a future where technology serves humanity, rather than deepening its divides.

Humanity now lives in an era of information abundance. Massive amounts of data are collected from every corner of the world, yet many fail to grasp how this affects their autonomy and the very essence of their lives. In just a short time, we've grown accustomed to a world where every mouse clicks, every second of video viewed, every message sent, every place visited, every device owned—even our smartwatches and home appliances—creates digital fragments of our existence. These fragments are collected and analyzed by a small circle of corporations, who now know us better than we know ourselves.

This technological shift has transformed us into objects of data extraction, reducing our humanity to fuel for the digital industrial revolution. But this phenomenon isn't new; its roots run deep into our history, tracing back to the Westphalian system of 1648, which established the foundations of modern international relations. The system centralized power in nation-states with the sovereign right to self-determination and fixed borders. While transformative for its time, these hierarchical principles remain largely unchanged, even in the face of today's digital realities. The centralized institutions that emerged over 400 years ago are still fundamentally intact.

Consider the principles guiding our modern parliaments and institutions. Despite the changing faces of leaders, the super-concentration of power persists, together with our unchecked biological appetites for domination. It perpetuates cycles of "more abuses, less expression of will," and "more power to the rich, fewer rights for the poor." In the digital age, this gap has only deepened.

This paradox defines our era: while we hold the technologies of the 21st century, our rights remain confined to the dusty pages of the Universal Declaration of Human Rights, written in the mid-20th century, and our institutions are stuck in an era closer to the 17th century.

Some innovators in the Web 3.0 space propose a bold solution: the emergence of network states—decentralized systems of governance that could redefine global power dynamics. Yet even these ambitious visions often overlook two critical elements:

1. *The Legal Foundation.* Web 3.0 developers, focused on technological innovation, frequently neglect the legal aspects essential for ensuring justice and equity. Without a robust legislative foundation and the recognition of digital rights, these new systems risk reproducing the same injustices and inequalities of the past. The creation of network states and start-up cities must integrate a normative framework—a horizontal, networked social contract—to ensure their governance is rooted in fairness and inclusivity.
2. *The Human-Centered Approach.* In designing the institutions of the future, we must prioritize the individual and their digital rights. The person must not become merely another link in a digital machine or a node in a data network. Governance systems must guarantee that individuals are treated as ends in themselves, not as abstract data points or expendable digital resources.

This book shifts the focus to the individual, particularly the youth, who is born into a world surrounded by digital technologies and becomes a "digital native" from the start. While many efforts have been made to adapt children to this digital society, these attempts often lean more toward restriction than enrichment.

According to UNESCO, one in six countries has banned the use of smartphones in schools. In China, authorities propose banning children from accessing the Internet on mobile devices between 10 p.m. and 6 a.m., while limiting online access for those aged 16 to 18 to just two hours per day. Meanwhile, platforms like YouTube prohibit users under 13 from registering. These measures reflect growing global concern about children's **digital rights** and the risks of **technology addiction**.

In the digital era, children are not just passive participants but **key stakeholders**. They represent the future and must become the central focus of human-centered systems of **governance, law, and education**. The digital age offers a unique opportunity to build systems that empower children, protect their rights, and prepare them to thrive as active, conscious participants in the world they will inherit.

This book argues that the cornerstone of a fair digital society lies in acknowledging and addressing the specific challenges faced by younger generations. They are not just participants in the digital world—they are its purpose and its future.

Before delving further, we must first understand how we arrived at this critical juncture and the interdisciplinary, philosophical nature of the underlying problem. Philosophy, long a beacon for answers to humanity's most profound questions, is intricately tied to social change and often serves as its catalyst. Yet, at times, philosophy falters, hesitating at the threshold of midnight, like the foreboding hands of a "doomsday" clock.

The foundational theoretical frameworks—or metanarratives—do not always evolve at the necessary pace. Society has transitioned from *modernism* to *postmodernism*, and now to *metamodernism*, which often casts shadows of "dark philosophy" and speculative theories.

- *Modernism* was an era of synchronization with time, where humanity created and directed the trajectory of progress, taking a central role in the world. Figures like Bacon, Descartes, Hobbes, and Rousseau embodied this period's optimistic belief in human agency and rationality.
- *Postmodernism*, on the other hand, dismantled the idea of a central narrative, dispersing meaning across the "plateau" of reality. It displaced time and modernity, relinquishing control over the course of history and progress. Thinkers such as

Deleuze, Foucault, McLuhan, and Baudrillard revealed a fragmented, decentralized world.

Both modernism and postmodernism, in their own ways, struggled with the concept of **time**. Today, we stand at the precipice of a temporal rift in the *socio-techno-natural physics* of global development, where the rapid evolution of technology collides with the inertia of social structures.

Contemporary theoretical lenses offer three dominant perspectives on this rift: *transhumanism, antihumanism, and posthumanism*, each reflecting a distinct path for navigating digital transformation. Alongside these, technocratic practitioners add layers *of techno-optimism* (championed by figures like Sam Altman) *and techno-pessimism* (articulated by Eliezer Yudkowsky). These frameworks shape the philosophical and practical debates on humanity's future:

1. *Transhumanism.*
   Transhumanism holds that the human self remains the central force in digital transformation. It envisions humanity enhanced by technology, retaining its privileged position as the primary driver of progress. This perspective, popular in Silicon Valley, is marked by aspirations such as "human enhancement," "life extension," "artificial life," and "the singularity." Transhumanism embraces technological innovation as a tool to transcend biological limits, empowering individuals to evolve alongside their creations.
2. *Antihumanism.*
   Antihumanism shifts away from humanity's position of uniqueness, positing that in the era of digital transformation, humans have ceded their central role to new actors, such as artificial intelligence. This view sees humanity as a **resource**—a set of data to be utilized or surpassed. Creativity in science, art, and influence is no longer exclusive to humans but is instead a relic of the past. Indicators of this worldview include concepts like the "rights of robots," "post-human society," and "autonomous AI," where human agency becomes secondary to machine-driven autonomy and progress.
3. *Posthumanism.*
   Posthumanism, in contrast to transhumanism and antihumanism, proposes a *horizontal symmetry* among actors, advocating for the coexistence of both human and non-human participants. This framework embraces hybridity, multi-

speciesism, and symbiotic relationships between humans and technology. Posthumanism integrates shades of techno-optimism and techno-pessimism, envisioning a vertical, networked space where all actors—human and non-human—have influence and agency. Its markers include concepts such as "cyborgs," "biotechnological mixing," and the "symbiosis of technology and humanity."

These three approaches to navigating the future echo the "isms" of past eras, rooted in times when information about the world was monopolized by the few. In those days, knowledge was taken on faith, and the right to access it was a privilege, not a universal right.

Today, these monopolies on knowledge have collapsed. Information is no longer the preserve of the elite—it is universally accessible. In this era of unprecedented access, we cannot afford to repeat the mistakes of the past. We must integrate scientific achievements, not just from the exact sciences but also from the humanities, to chart a path forward.

By acknowledging the interplay between these philosophical frameworks and the technological realities of our time, we can begin to understand the complexities of this pivotal moment. Our task is to transcend outdated ideological directions and create a more inclusive, human-centered narrative for the future.

Future startup cities, network states, and global governance institutions are poised to become new benchmarks for social organization. They demand a reimagining of governance and coordination—grounded in decentralization, scientific orientation, openness, flexibility, and globalization. In this evolving context, philosophical frameworks like humanism, antihumanism, and posthumanism offer valuable tools for conceptualizing how to structure this new reality—or serve as warnings of how not to. A genuine understanding of how to create a world of digital prosperity, rooted in robust legal norms and a global scientific method tailored for the 21st century, is essential.

In this endeavor, artificial intelligence (AI) can play a pivotal role. The future philosophy of AI, I believe, must include a redefined Turing test—one that evaluates whether AI models align with global interests of sustainable development and adhere to human and digital rights, with particular emphasis on protecting and empowering children and young people.

However, debates about temporary moratoriums and a "human-centric" approach based on corporate self-regulation are inadequate, offering only superficial and short-term solutions. For instance, Elon Musk, while vocally supporting a pause in AI research, simultaneously launched his own AI-focused company, xAI ("Understand the Universe"). This performative contradiction highlights the growing divide between the rhetoric and actions of global digital leaders. Such moves can be interpreted as strategic attempts to outpace competitors in the race toward AGI (Artificial General Intelligence) while preserving markets for their products—an approach that reflects the underlying philosophy of eternal scaling embedded in contemporary business models.

These actions expose a deeper gap between political-legal realities and economic-digital ambitions. Bridging this divide requires not just new governance frameworks but a shift in collective priorities, ensuring that technological advancements benefit all of humanity rather than exacerbating inequalities.

This book seeks to contribute to the creation of a new conceptual narrative framework that enables the effective functioning of digital society while addressing the gaps. By embedding global digital human and child rights into this framework, we aim to establish a unified theoretical foundation for building a sustainable civilization.

The stakes could not be higher. Divergent viewpoints—like planets scattered across the map of a cosmic atlas—will inevitably lead to the formation of vastly different societies, states, and systems of global governance. Without a shared vision, humanity risks creating a future defined by millennial inequality and the realization of dystopian scenarios on an exponential scale.

This book accelerates the search for a collaborative global strategy—one that fosters prosperity, creativity, and sustainable development, and unites humanity in shared purpose. Only through such collaboration can we ensure that the promise of our digital age leads to a thriving and equitable civilization for all.

The methodology presented in this book integrates the perspectives of seven key stakeholders: **the public sector, private sector, technical sector, legal sector, educational sector, media sector, and representatives of the global community.** By blending, refracting, and stripping their actions of digital risks, it aims to develop practical tools, concepts, and frameworks to help these stakeholders adapt to the era of decentralized network development.

The lessons of past industrial revolutions remind us of the critical importance of conscious decision-making in times of technological upheaval. It is no longer enough to assume that new technologies inherently lead to universal prosperity. The question of whether they serve the public good has transcended technical considerations to become an issue of personal agency, law, ethics, governance, and politics.

Technology—be it the omnipresent Internet or the pervasive influence of artificial intelligence—is not merely shaping our tools. It is reweaving and reforging the very fabric of our social structures, norms, cultures, and the societies in which we live. This dynamic transformation creates an ever-expanding field of social interaction, political dialogue, and governance. At the same time, it imposes an urgent need to bring the humanities to the forefront of modern progress.

In an age when technological development no longer accompanies social progress but often dictates it, the humanities act as our guiding light, preserving the essence of humanity amidst the vast and all-consuming grandeur of technology.

In response to these challenges, this book develops a human-centered legal approach to guide managed global sustainable development. It introduces the foundation for a digital sustainability framework for children, spanning their journey from receiving their first digital footprint—an ultrasound image—to reaching adulthood in a rapidly changing digital world[26].

Our approach aligns with the "Youth 2030 Strategy," embracing its rhythm, no matter how difficult or unpredictable the journey. As the UN Secretary-General has emphasized, youth participation must become the norm, influencing decisions, policies, and investments. The establishment of the Youth Office within the UN Secretariat serves as a critical milestone, ensuring protection, coordination, and accountability for young people. In this context, adapting the digital space to meet the needs and aspirations of youth becomes an essential challenge of the digital transformation era.

The current generation is the key to shaping the future and must therefore be equipped with innovative tools, a catalog of digital rights,

---

[26]Burianov M.S. Prospects for international legal consolidation of the digital rights of the child // Actual problems of economics, management and law. Actual problems of economics, management and law / ed. Yu.V. Gavrilova. Saratov: Publishing house "Saratovsky Istochnik", 2023. 1164 p. – P. 257–261.

and active involvement in the creation of future digital network institutions.

This book does not claim to offer universal ethical solutions. Instead, it serves as an invitation to reflect and provides a set of tools to escape what Yuval Noah Harari described at the World Economic Forum as a state of "philosophical bankruptcy" in addressing digital challenges. These challenges influence law, politics, and society over the long term.

Simultaneously, this book charts a path for every individual to participate actively in the process of digital transformation. Such a transformation, which causes tectonic shifts across all areas of social relations—not just in the economic sphere—is too significant for any of us to remain mere spectators. The future demands action, collaboration, and thoughtful engagement from all of humanity.

INTRODUCTION: TECHNOLOGIES AS AN EMBODIMENT OF INDIVIDUALITY

The steady hum of the factory floor has given way to the silent, tireless flow of generative AI.

This new force does more than replicate products; it creates experiences shaped to each person.

The bleak futures once imagined in fiction—of mass unemployment, chaos, and lost control—are giving way to a different reality: a dynamic landscape designed around the contours of our individuality.

Here, uniqueness is the most valuable currency.

Our age, culture, vulnerabilities, and strengths shape every step of our digital path.

Medicine adapts to our biology.

Assistants learn our habits.

AI therapists respond to our moods.

Education bends to our needs.

Technology no longer erases the individual; it amplifies it.

It invites us to weave authentic digital identities—threads that link our distinctiveness into an inclusive global tapestry.
Let us enter this generative age not as passive recipients, but as active creators—participants in a future shaped by uniqueness, collaboration, and shared purpose.
The magic of technology lies not in the machines themselves, but in how they expand our sense of self, opening new dimensions of meaning, creativity, and connection.

Yet this promise must rest on a human-centered legal foundation.
Ethical guidelines, however sincere, are no longer enough.
In the era of Industry 4.0, legal frameworks must prioritise the realisation of human rights over mere protection.

They must counter the sameness of echo chambers, protect the vast landscapes of our personal data, and ensure that as we reveal our individuality to machines, it is never misused. Equally, accessibility must be universal—across borders, cultures, and abilities—so that no one is left behind.

Artificial intelligence can free us from repetition and routine, unlocking time and energy for creativity, innovation, and deeper purpose.

Its role is not to replace humanity, but to enhance it—building a digital world where innovation serves dignity, inclusivity, and opportunity.

As we step into the digital horizon, we must ensure this transformation is one for humanity, by humanity, and about humanity—where rights, identities, and uniqueness form the cornerstone of a just and sustainable digital age.

Generative AI, social media, and information and communication technologies are now woven into children's daily lives, shaping how their rights are realised. They offer opportunities for learning, connection, and self-expression, but also bring risks—unlawful data collection, online exploitation, algorithmic bias, and threats from digital surveillance and militarisation.

In today's connected world, a child's digital footprint can begin before birth, with ultrasound images or birth announcements shared online.

From there, personal details—birthdays, names, locations—can be captured by data brokers and sold to advertisers. Research shows that more than 80% of children have an online presence by the age of two.

The pervasive nature of technology necessitates the acquisition of digital skills, directly impacting children's learning and development. Continuous internet connectivity can render children vulnerable to rights violations. Many parents are not fully aware of their children's online activities or the associated risks, highlighting the need for enhanced digital literacy among adults. A survey revealed that approximately 29% of parents are unaware of the information about their child that is publicly available online, and 62%[27], do not seek their children's permission before sharing content about them.

---

[27]All-in-one app to protect kids in the digital world / URL:https://www.kaspersky.ru/safe-kids(date accessed: 26.07.2023).

Addressing these challenges requires a comprehensive approach that emphasizes the realization of children's rights in the digital realm. This includes implementing robust legal frameworks that extend to Industry 4.0 technologies, ensuring ethical and lawful regulation, and promoting digital literacy among both children and adults to navigate the complexities of the digital age effectively.

Technological progress consistently outpaces legislation, particularly on a global scale where development is unevenly distributed. The most pressing challenge lies in safeguarding vulnerable groups, especially children. To address this, we must take on the role of digital architects of our shared future, crafting systems and laws that not only protect but also empower children and society against the myriad potential threats in this fast-evolving digital landscape. It is essential that parents, educators, and policymakers rely on effective tools and internationally recognized principles to ensure children's safety, agency, and development in the digital environment.

Looking ahead, technology holds the potential to raise the well-being of societies, especially for future generations.

As Sam Altman suggests, we may need a "Moore's Law for everything," making essentials like housing, education, food, and clothing exponentially more affordable over time.

To this, we can add a "**Moore's Law for human rights**"[28]: a vision where technological progress is matched by fair frameworks that distribute its benefits universally, starting in childhood.

This approach aligns with the Sustainable Development Goals, linking economic progress directly to social progress.

Our shared future can be a mosaic of diverse human experiences, each piece carrying intrinsic value.

A truly inclusive world respects differences in age, background, and circumstance, ensuring that every voice is heard, every value recognised, and every right protected.

---

[28] Moore's Law for Everything/ URL:https://moores.samaltman.com/(date accessed: 01.12.2024). Note: the classic Moore's law states that the basic characteristics of computers improve twofold every two years. Note that Altman sponsors a large-scale study on unconditional basic income (UBI). More details: https://youtu.be/L_Guz73e6fw?t=6017.

Such a society does not merely adapt to change—it thrives by grounding innovation in dignity, equity, and human potential.

CHAPTER 1. REIMAGINING HUMAN'S AND CHILDREN'S RIGHTS IN THE DIGITAL AGE

1.1. The Evolving Concept of Human's and Children's Rights in a Digital World

Understanding children's rights depends on how we see the child and the fast-changing world they inhabit. As the foundation of human-centred legal regulation, these concepts are essential to closing the gap between outdated laws and the rapid pace of societal and technological change. This gap is most visible in the absence of coherent frameworks for children's digital rights—frameworks that must ensure equitable access to the economic and digital resources of a hyperconnected world while addressing unprecedented risks.

In the 21st century, children grow up in an environment that is vastly different from the analog world in which traditional human rights frameworks were conceived. The environment in which a child grows has changed dramatically compared to the moment when human and child rights were consolidated in the period from the 50s to the 80s of the 20th centuries, tied to the transition from the second industrial revolution to the third, from mechanical engineering to programming. Today's reality is a hybrid of physical and digital dimensions, shaped by artificial intelligence, big data, neural interfaces, and pervasive connectivity like GPS. These technologies offer children unparalleled opportunities for personalized education, adaptive healthcare, and meaningful global interactions, but they also expose them to significant online threats and raise questions about the impact of digital environments on their cognitive, emotional, and social development.

As we embark on this research, it is vital to reflect on the seismic changes that have unfolded over the past ***century***. Philosopher Michel Serres captured this transformation poignantly, observing, "Our culture, previously tied to agriculture, has suddenly shifted... Children today no longer inhabit the land as before; they engage with the world through a digital lens. Their existence is defined not by pastoral landscapes but by screens, where rapid images and fragmented narratives dominate their reality. They live in an overcrowded world of seven billion souls, and their history, shaped by mass media, is one of distraction, brevity, and relentless exposure to death."[29]

Indeed, Serres' insights underline the urgency of reassessing children's rights in this new era. The internet and digital ecosystems have fundamentally altered how children think, learn, and engage with the world. They process information differently, navigating virtual landscapes where proximity is defined not by physical distance but by the instant availability of knowledge, connections, and experiences.

Following this perspective, we delve deeper into how these changes reshape children's rights. As Serres notes, "Children today inhabit a topological space where everything is adjacent to everything. With mobile phones, they can reach anyone; with GPS, they can locate anywhere; with the internet, they have access to an infinite repository of knowledge." This digital landscape has expanded their horizons but also necessitates the development of new skills, such as critical thinking and emotional resilience, to navigate a world where artificial intelligence augments memory but risks diminishing awareness and attentiveness.

The integration of AI further complicates this dynamic. Generative AI systems, with their vast capacity to analyze and synthesize knowledge, offer answers to virtually any question, transforming how children learn and interact with information. However, this also shifts the developmental focus from information retention to higher-order skills like reflection, ethical reasoning, and adaptability. These shifts challenge us to reimagine the rights of children in a way that embraces the benefits of technology while safeguarding their humanity.

By exploring these questions, this chapter sets the stage for a deeper investigation into the principles and legal frameworks needed to ensure that children's rights evolve in tandem with technological progress. The aim is not just to protect, but to empower children as active participants in shaping their digital futures.

Before addressing the specific challenges and opportunities of children's rights in the digital age, it is important to place them within the broader historical and philosophical context of human rights. Understanding how the very concept of rights has evolved—and how it frames the relationship between the individual, the state, and society—provides a foundation for redefining protections and freedoms in an era of rapid technological change.

---

[29]Serr M. Little Thumb. - Moscow: Ad Marginem Press, sor. 2016. - 77 p. - P. 5.

Understanding human rights

Understanding human rights begins with understanding what it means to be human in a given historical and social context. Legal scholars define human rights as the most essential capacities for personal development—determining the scope of one's freedom and the relationship between the individual and the state. They encompass political, social, cultural, and economic domains, and serve as a measure of the democratic character of any political-legal system. The position of human rights in the hierarchy of legal values shapes not only the interaction between the individual and the state, but also between humanity and nature, and between the citizen and the global community.

The modern recognition of human dignity as inviolable emerged only after the devastation of two world wars. This turning point required a rebalancing of global governance so that the power and ambitions of individuals—especially those with privileged access to authority and technology—would be constrained by a renewed legal status for the person. The intellectual lineage of this idea is deep: Jean-Jacques Burlamaqui saw rights as the most direct path to achieving one's goals; René Cassin, a principal architect of the Universal Declaration of Human Rights, framed human rights as a field of knowledge dedicated to human dignity and the capacities needed for the full development of each person; and 19th-century philosopher Bruno Bauer argued that rights are not a gift of nature or history, but the result of struggles against inherited privilege.

Human rights are thus both a shield against discrimination and a foundation for a society based on mutual respect, regardless of gender, race, worldview, place of birth, age, or other markers of human diversity. Their realisation is also a test of our education systems and of our ability, as a global community, to adapt to new realities. UNESCO has long recommended human rights education as a core subject in schools worldwide.

In contemporary international law, Karel Vasak's concept of the "three generations" of human rights—corresponding to liberty, equality, and fraternity—remains influential. Yet the accelerating pace of technological change suggests that a fourth generation may now be emerging, one that recognises the need to safeguard rights in the digital sphere as rigorously as in the physical world.

This historical perspective makes clear that children's rights cannot be treated as an isolated category. They are part of the broader evolution of human rights, yet they face distinct pressures in the digital era. As the boundaries between physical and virtual environments blur, and as children spend more of their formative years in online spaces, the principles forged in earlier centuries must be reinterpreted to safeguard dignity, equity, and opportunity for the youngest members of our global society.

The Child in the Digital Age: A New Paradigm for Rights and Development

By the end of the 20th century, the foundational elements of children's rights were consolidated during the industrial era, but they largely neglected the needs of the emerging technological and digital age. Today, children inhabit a predominantly digital world. With one-third of all internet users being underage, their early social and emotional experiences are increasingly shaped by online environments—social networks, video-sharing platforms, and immersive online games. These digital spaces act as modern arenas for exploration, self-expression, and the initial formation of identity.

While society has spent over a century weaving itself into the fabric of the digital era, we are now standing on the threshold of a new frontier. This coming wave of technological evolution, driven by creative artificial intelligence, quantum computing, and immersive digital realities, is poised to engage more of our senses, expand our perceptions, and redefine the essence of human creativity, individuality, and dignity—all of which are enshrined in human and child rights[30].

In today's era, progress is inseparable from our ability to harness data and digital technologies. These advances challenge us to reconsider the age-old question: *"What does it mean to be human in the modern world?"* This question has taken on new dimensions in a landscape where creative tasks are increasingly executed by AI systems, intellectual exploration is aided by machine-learning chatbots[31], and programs analyzing biometrics and voice patterns can glean insights about individuals faster than those individuals may ever understand

---

[30]Danilova V.A. Children's rights as a national priority of the Russian Federation in the context of digital transformation of society // Law and state: theory and practice. 2022. No. 6 (210). - P. 16.

[31]Pashnina T. V. Trends in the implementation of information rights of minors in the library sphere in the context of digital transformation // Bulletin of the O. E. Kutafin University. 2022. No. 4 (92). - P. 92.

themselves. The rapid acceleration of history through technological microchanges has created a cumulative, tectonic shift in societal structures and global dynamics.

Among all groups, minors are the most vulnerable in this evolving digital and informational context. This compels us to ask an equally pressing question: *"What does it mean to be a child in today's world?"* What opportunities, protections, and rights should children have to ensure their full self-realization and development in this era of digital transformation?

The Infosphere and the Rise of "Information Organisms". Technology is not just shaping human behavior—it is fundamentally altering how we think, interact, and perceive reality. Luciano Floridi, Professor of Philosophy at Oxford University, introduced the concept of "**Information Organisms**" (short for information organisms) to describe modern humans and children as entities intertwined with the infosphere—a global digital ecosystem where algorithms and data flows redefine human experience[32].

Floridi describes Information Organisms as *"information organisms that exist in symbiosis with biological agents and engineering artifacts within a global environment—the infosphere."* This concept recognizes that we no longer exist solely as biological beings. We are increasingly integrated with data streams, algorithmic systems, and interconnected technologies that augment, and sometimes challenge, traditional forms of human interaction and identity.

In this context, children are not merely participants in the digital world—they are being shaped by it in profound ways. Algorithms curate their experiences, platforms influence their thinking, and data flows increasingly define their opportunities. As a society, we must critically assess how these dynamics affect the rights and development of children in the digital age.

Towards a New Definition of Child Rights in the Digital Era

To navigate these complexities, it is imperative to redefine child rights for the digital age. This means not only protecting children from online harms but empowering them to thrive in a world shaped by

---

[32]Information: A Very Short Introduction (Very Short Introductions) Illustrated Edition // URL:https://www.amazon.com/Information-Short-Introduction-Luciano-Floridi/dp/0199551375(date accessed: 25.05.2023).

artificial intelligence, big data, and quantum technologies. The opportunities for adaptive learning, creative expression, and global connectivity are boundless, but they must be balanced with ethical and human-centered safeguards.

As we confront these challenges, we must also recognize the unique creative and intellectual potential of the younger generation. They are not only passive recipients of technology but active participants in shaping its future. By prioritizing their rights and ensuring equitable access to the tools and benefits of the digital era, we can empower children to become architects of a more inclusive and sustainable technological future.

This chapter lays the foundation for understanding how the digital revolution is redefining childhood and highlights the urgency of adapting our legal, ethical, and societal frameworks to meet the needs of this new generation of "inforgs."

The Child in the Infosphere: A New Context for Rights and Identity

The *infosphere*, as described by Luciano Floridi, is "an information environment constituted by information processes, services, and objects, including information agents with their properties, interactions, and mutual connections." This interconnected and data-driven ecosystem gives rise to what can be termed *information law*—a legal framework essential for navigating the complexities of the digital age[33][34].

When we discuss children within this context, we consider them not only as individuals of today but as pioneers of a rapidly unfolding future. Global, uncontrolled digitalization is already shaping their development in profound and unpredictable ways. Scientific studies have demonstrated that digital environments can significantly alter the properties of adult personalities, raising an urgent question[35]: *What*

---

[33] Floridi, L. Philosophy and Computing: An introduction. Routledge. 1999. – 257 p.

[34]Rassolov, I. M. Information law: textbook and practical training for universities / I. M. Rassolov. - 6th ed., revised and enlarged. - Moscow: Yurait Publishing House, 2023. - 415 p.

[35]Stieger, M., Nißen, M., Rüegger, D. et al. PEACH, a smartphone- and conversational agent-based coaching intervention for intentional personality

happens when the digital environment becomes a second womb for children's development? Their digital identities, formed in this overlapping real and virtual world, could predetermine their growth in both realms, creating a reality where the digital ecosystem is inseparable from human existence.

Children in the Web of Opportunities and Risks

In the era of universal connectivity, children are both empowered by a web of digital opportunities and ensnared by its inherent risks. The digital environment creates conditions that not only shape their cognitive abilities but also embed values, ideas, and identity markers within their personalities. This dynamic interplay is increasingly manipulated by external forces, whether authoritarian states imposing ideological conformity or commercial entities driven by profit motives.

This phenomenon echoes the famous domestication experiment conducted by academician Dmitry Belyaev, where foxes were selectively bred to alter their behaviors and traits. Similarly, digital applications—through targeted algorithms, patterns of thinking, and social programming—are capable of influencing children's developing minds in an accelerated manner. They introduce notions of "good" and "bad," "us" and "them," directly invading the nascent identity of young individuals. Yet, unlike Belyaev's foxes, the consequences of such interventions in the human psyche, particularly when applied to children, are not yet fully understood.

This intervention, this invisible hand that can sculpt or distort a child's intellectual and emotional development, calls for vigilance. It demands a collective responsibility to shield not only the present but also the uncertain future of our children, ensuring that their rights are safeguarded against the unpredictable consequences of unchecked technological influence.

Defining Children's Rights in the Digital Age

Amidst the tidal wave of technological changes, providing a precise definition of *children's rights* becomes crucial. These rights represent the opportunities afforded to every individual under 18 years

---

change: study protocol of a randomized, wait-list controlled trial. BMC Psychol 6, 43. 2018. – 15 p.

of age for their full development, as enshrined in both international and domestic law.

The concept of a *child* was first legally defined in Article 1 of the UN Convention on the Rights of the Child and echoed in Paragraph 1, Article 54, of the Russian Federation Family Code: *A child is any human being below the age of 18.* These rights remain valid until the individual reaches adulthood. However, this definition, formed in the late 20th century, predates the digital revolution and does not yet fully encapsulate the unique challenges and opportunities presented by today's digital world.

A New Paradigm for Child Rights in the Infosphere.

As the infosphere continues to evolve, the rights of children must be reframed to address this new reality. The digital ecosystem, which now serves as an extension of the child's developmental environment, demands legal frameworks that not only protect against exploitation and harm but also actively foster opportunities for growth and self-expression.

This new paradigm must ensure that children are not treated as passive subjects of technological progress but as active participants with rights and agency in shaping their digital identities. By embedding these rights within the foundations of international and domestic law, we can empower the youngest members of society to thrive in a world where the digital and the physical are increasingly intertwined.

Guardianship and trusteeship authorities, the prosecutor, the court, and the minor himself are obligated to protect the rights of the child if their interest's conflict with those of the parents. The legal status of a child encompasses their rights, duties, and responsibilities from birth until reaching adulthood. The legal representatives of a minor child in the exercise of their rights are the parents or individuals acting in their place[36][37].

The rights of the child are recognized as a universal human value. Childhood, as a unique and unrepeatable period in the development of human personality, underscores the importance of considering the rights

---

[36]Family law: protection of rights and legitimate interests of minor children: textbook / [T. L. Kalacheva et al.; scientific ed. N. S. Makharadze]. - Khabarovsk: Publishing house of Pacific state University, 2019. - 118 p. - P. 20.

[37]Children's Rights. Reference publication Samara 2017 / URL:https://op63.ru/attachments/Prava%20detej%202017.pdf(date accessed: 25.05.2023).

of the child as a fundamental value. The future of humanity and individual states is intrinsically linked to the cultural and physical development of the younger generation and their understanding of the role and significance of human rights within society.

At the same time, the rights of the child and human rights form an inseparable whole. Human rights serve as a foundational framework that ensures the comprehensive protection and fulfillment of children's rights.

The rights of the child are shaped by the unique characteristics of their living conditions and developmental needs, which include: 1) age; 2) dependence on parents or guardians; 3) psychological and emotional vulnerability; and 4) distinct cognitive processes influenced by neurogenesis and the inherent openness of children to learning and experience.

For children, as a particularly vulnerable social group, two categories of rights are especially critical.

Personal Rights:

These encompass the fundamental rights that safeguard a child's individuality and well-being:

1. The right to life and health;
2. The right to a name and identity;
3. The right to live within a family environment;
4. The right to happiness;
5. The right to a safe existence, free from exploitation, neglect, or abuse;
6. The right to freely express opinions;
7. The right to protection from cruelty, bullying, and unfair treatment;
8. The right to citizenship and associated protections.

These inalienable rights are indispensable for fostering the child's growth into a healthy and independent individual.

Socio-Cultural Rights:

These rights enable the holistic development of a child and include:

1. The right to comprehensive development—intellectual, psychological, physical, and emotional—while ensuring a decent standard of living;
2. The right to quality education;
3. The right to health and well-being;
4. The right to access cultural values and participate in cultural life.

Children's rights are aptly described in legal scholarship as "norms enshrined in legislation to fulfill a child's basic needs—survival, development, socialization, and assistance in crisis situations that children may encounter."

The *Large Law Dictionary* defines children's rights as "human rights applied specifically to children, justified by the fact that children, unlike adults, lack a full set of legal capacities and yet require distinct protections related to their age, role within the family, and unique developmental needs."

Similarly, an English legal dictionary defines children's rights as the entitlements that allow children to participate in decisions affecting their lives, as well as the rights to live free from hunger, abuse, neglect, and other inhumane conditions. These rights form the foundation for protecting and empowering children, ensuring they thrive in both personal and societal dimensions.

The rights of the child, as defined in the dissertation of V.A. Abramov, are conceptualized as human rights specifically applied to children, considering their age-specific characteristic of being minors. This perspective emphasizes that children's rights inherit the essential attributes of human rights while accommodating the unique developmental needs of youth.

Contrary to the interpretation presented in N.I. Eliasberg's *Primary School Textbook*[38], which posits that "rights are what society and the state allow a person," this understanding risks reducing innate rights to mere permissions granted by external entities. Children's rights are intrinsic by birth and not contingent on societal or governmental approval, avoiding the potential conflation of rights with duties.

As noted in a methodological publication, public and professional discourse on children's rights often elicits a polarizing response: "You speak of rights, but why not responsibilities?" This reaction reflects a gap in legal literacy and an outdated view of children as beings not yet equal to adults. This is particularly misplaced considering the 1989 UN Convention on the Rights of the Child, which firmly established the legal status of children as rights holders. Rights, as conceptualized, represent

[38]Eliasberg N. I. The rights of the child are your rights! A teaching aid for primary school. Editorial board: N. I. Eliasberg (chairman), with the participation of R. G. Khusainov and O. E. Semenova (selection of materials for the book). - St. Petersburg: Publishing house "Tree of Life", 2011. - 64 p. - P. 10.

a framework for permitted, conscious behavior that does not harm others, underscoring freedom and agency even at a young age.

In summation, the rights of the child can be defined as a set of essential opportunities afforded to every individual under 18, designed for their preservation and holistic development, and rooted in international and domestic law. These rights are an extension of universal human rights, encompassing key protections and freedoms:

1. The right to life and healthy development;
2. The right to a name, citizenship, and care by one's parents;
3. The right to individuality and personal identity;
4. The right to family unity, unless separation is in the child's best interest;
5. The right to cross borders to maintain family ties;
6. The right to independent views, irrespective of age;
7. The right to freely express opinions;
8. The right to freedom of conscience and belief;
9. The right to freedom of association and peaceful assembly, within safety guidelines;
10. The right to privacy and protection from arbitrary interference;
11. The right to access information fostering moral, spiritual, physical, and psychological well-being.

The Potential and Risks of Digital Technologies for Children

Digital technologies have a profound impact on children's cognitive development, offering both opportunities and risks. As explored in studies on the effects of premature and excessive use of these technologies, such as the insights presented in Digitale Demenz (2012), the unregulated and ineffective use of digital tools can lead to a decline in cognitive functions. These effects, particularly on developing brains, may be long-lasting or even irreversible. However, alongside these potential dangers lies an array of transformative opportunities, particularly in the realm of education and personal development.

Human history demonstrates how new tools reshape our abilities and expand our potential. For example, the invention of writing fundamentally altered memory functions, allowing humanity to record, share, and build upon knowledge in unprecedented ways. Similarly, today's technologies—generative artificial intelligence, virtual environments, and the metaverse—offer children unique pathways to learn, interact, and grow. These innovations can support the implementation of children's rights in a tailored and meaningful way, accounting for their individuality, age, gender, maturity, and socio-cultural context.

To harness these opportunities while mitigating risks, it is essential to develop a balanced approach to technology. This involves not only advancing digital literacy but also fostering a critical awareness in both children and adults. As we navigate this digital age, our role as educators, policymakers, and parents is to guide children in using these tools wisely and consciously. By doing so, we can ensure that digital technologies empower children rather than diminish their potential.

The evolving digital ecosystem must reflect the rights of children, enabling them to flourish in a personalized and inclusive environment. Whether through adaptive learning systems, enhanced access to information, or innovative healthcare solutions, technologies must be leveraged to support the holistic development of each child, while safeguarding their fundamental rights.

Box 3. Legal Understanding as the Key to Governing Digital Technologies and Contemporary Society — Essential Concepts

Human rights — the inherent capacities of every person, from birth, for human existence and development; enshrined in international law and domestic legislation; an expression of human dignity realized through effective public institutions and the rule of law; a single, integrated set of entitlements that is indivisible, interrelated and interdependent.

Children's rights — the inherent entitlements of every person under 18 to survival, development, protection and participation, secured in international and domestic law; distinguished by the primacy of the child's best interests and recognition of evolving capacities, requiring special safeguards and proactive measures by duty-bearers. This book prioritizes the youngest cohorts most exposed to asymmetric power and risk in digital environments.

Duty-bearers in the digital age — responsibility for the realization and protection of rights lies with both *States and corporations*, as major technology companies now shape freedoms and choices at scale.

Digital commons — treatment of key platforms as part of a shared public realm, orienting governance to human rights and the public interest rather than solely to profit.

Limits on data extraction and manipulation — strict constraints on data collection and behavioral engineering by platforms, with explicit priority to users' rights, especially those of vulnerable groups.

Reframing the AI race — from competition for power and profit to a race for shared prosperity and expanded human capabilities for all.

In the next section, we will explore how modern global processes and Industry 4.0 technologies intersect with children's rights and the law, shaping the future of their digital and real-world experiences.

1.2. Global Digital Revolution: Industry 4.0, Web 3.0, and Their Influence on Children's Rights

Children are active users of the Internet and mobile technologies, making up one in three Internet users worldwide. For many children, the line between the online and offline worlds is blurred.[39].

In our rapidly changing world, where technology erases the boundaries of space and time, the Internet is becoming a global phenomenon and a prerequisite for responding to the challenges of the 21st century. In 2022, 5.3 billion people (66% of the world's population) had access to the Internet. This represents a significant increase of 65% since 2015, when only 40% of the world's population was online. Globally, 69% of men and 63% of women use the Internet. This means that in 2022, there were 259 million more men than women using the Internet, according to a special SDG Progress Report.

Five interconnected trajectories of global digital transformation are set to shape the period from 2024 to 2030:

1. The Rise of Web 3.0
   We are entering a new era of digital space defined by decentralized communities and institutions. This shift introduces decentralized partnerships that fundamentally transform approaches to trust, sovereignty, transaction costs, and the concentration of power.
2. The Integration of Virtual Reality and Neural Interface Technologies
   A new world is emerging where the boundaries between physical and digital realities blur. As virtual reality spreads and neural interfaces advance, we are moving toward an era where human experience seamlessly spans both dimensions.
3. The Maturation of Fourth Industrial Revolution Technologies
   Technologies that once fueled nascent start-ups are now transitioning to established companies, scaling their impact. This marks a critical shift as these innovations become integral to industries worldwide, driving progress at an unprecedented pace.
4. The Exponential Growth of Applied Generative AI
   Generative artificial intelligence is on the verge of revolutionizing economies by drastically reducing the marginal cost of goods and services. This promises to make essential resources more accessible, democratizing benefits across societies and reshaping global markets.

---

[39]Livingstone, S., Carr, J. and Byrne, J. (2016). One in Three: Internet Governance and Children's Rights. Innocenti Discussion Paper No.2016-01, UNICEF Office of Research, Florence. URL:https://www.unicef-irc.org/publications/pdf/idp_2016_01.pdf(date accessed: 25.05.2023).

5. The Tech Revolution Across Traditional Industries Traditional sectors are increasingly incorporating innovative technologies, adding a "Tech" prefix to every domain: MedTech, EdTech, GovTech, and soon HumanTech. These advancements signal a new era of interdisciplinary innovation. For instance, neural interfaces may soon serve as additional RAM for the human brain, augmenting memory and cognition.

Internet Web 3.0. Today, we are witnessing the emergence of a transformative new phase in the evolution of the Internet—Web 3.0. This stage marks a significant leap forward in global technological innovation, reshaping how societies interact, create, and address the challenges of the modern era. As highlighted in the Human Development Report 2022, innovation—whether technological, economic, or cultural—is essential for navigating the unknown and complex challenges humanity faces. Yet, even as we participate in building the digital ecosystem of Industry 4.0, unresolved issues from earlier technological eras, such as Industry 3.0 and Web 2.0, continue to affect vulnerable groups, particularly adolescents.

To fully grasp the importance of Web 3.0, it's essential to revisit the milestones of Internet evolution:

The Genesis of the Internet: Web 1.0 and the Birth of a Digital Era. This initial phase of the Internet was largely a one-way communication platform, characterized by:

1. Limited interactivity and automation, making it a static space for information consumption.
2. Strict hierarchies, where webmasters-controlled content creation, and users primarily consumed information.
3. Minimal user participation, with users acting as passive observers rather than active contributors.
4. Simplistic website design, with limited functionality and aesthetic appeal.

Web 1.0 refers to the first version of the Internet, an era that traces its origins back to the Defense Advanced Research Projects Agency (DARPA). What began as a government initiative evolved into the World Wide Web—a digital communication platform that symbolized the future of connectivity. At its core, this early Internet was built on static hyperlinks leading to rudimentary web pages. These pages lacked the interactivity, sleek design, and advanced features that define the web we know today. Yet, this simplicity laid the foundation for a technological revolution.

A pivotal figure in this transformation was Vinton Cerf, often called one of the "fathers of the Internet." Cerf's pioneering work helped transition the Internet from the basic hyperlink-driven structure of Web 1.0 to the dynamic, user-centric platforms we see today. His contributions marked the beginning of a shift toward a more interconnected and interactive digital experience.

One of the most defining moments of the Web 1.0 era was the creation of the Google search engine in 1998. Google revolutionized how people accessed information, setting the stage for the data-driven digital economy that thrives today. It introduced new paradigms in online advertising and catalyzed the emergence of Internet startups, offering unprecedented opportunities for global self-expression and innovation. This milestone not only redefined how we navigate the Internet but also opened doors to boundless possibilities for personal and professional growth, transcending geographical and cultural barriers.

Web 2.0: The Internet Comes Alive. Web 2.0, the dominant phase of the modern Internet, transformed how users engage with the digital world. Unlike its static predecessor, Web 2.0 introduced a dynamic and participatory web, characterized by several key aspects:

1. The Rise of Social Networks. Platforms like Facebook, Twitter, and Instagram redefined how people connect, communicate, and share their lives online, embedding themselves into daily routines.
2. Corporate Dominance. Large tech corporations emerged as the architects of the digital landscape, setting trends and steering technological progress.
3. User-Generated Content. Unlike Web 1.0's passive consumption model, Web 2.0 empowered users to actively shape the Internet by creating and sharing content.
4. Centralized Data Storage. Information continued to be stored centrally on servers, delivered to users on request, enhancing accessibility.
5. Data as a Commodity. User data became the new oil—highly sought after by advertisers and corporations, driving a data-centric economy.

This phase of the Internet fostered interconnectedness, enabling users to share and interact with content more fluidly. Web 2.0 blurred the lines between creators and consumers, giving rise to a vibrant ecosystem where anyone could influence the global narrative.

The Era of the User.

The peak of Web 2.0's influence came in 2006 when *Time* magazine named "You" as Person of the Year. This landmark recognition celebrated the collective contributions of millions who created and shared content on platforms like Wikipedia, YouTube, and MySpace. It acknowledged the democratization of influence in the digital sphere, where the individual became a driving force behind innovation.

Fast forward to 2023, and *Time* placed ChatGPT on its cover, signaling the dawn of another transformative moment. As generative AI begins to reshape industries and daily life, the question arises: Will artificial intelligence ever be named Person of the Year?

The Future of Recognition in the Digital Age.

By 2025, it might not be far-fetched to see a ChatGPT user on *Time*'s cover—a nod to those who harness AI to amplify their creativity and analytical capabilities, reminding us that humans remain the architects of technological progress. Between 2027 and 2029, however, the spotlight could shift to AGI (Artificial General Intelligence), which may begin solving humanity's grandest challenges.

Will AGI revolutionize *Time*'s tradition by founding its own magazine, perhaps titled *Eternity*, to celebrate the achievements of the most advanced AI agents? The idea of AI honoring its own might feel like science fiction today, but in the fast-moving digital age, the future often arrives sooner than we expect.

Web 3.0: A New Era of the Internet. Web 3.0 heralds a revolutionary phase in the evolution of the World Wide Web, bringing with it transformative characteristics that set it apart from its predecessors:

1. Ubiquity. With the proliferation of IoT devices and smart gadgets, the Internet in the Web 3.0 era aims to be omnipresent, seamlessly integrating into every aspect of life.
2. Freedom. Community moderation replaces corporate oversight, enabling a more democratized form of content control.
3. Openness. The widespread adoption of open-source software fosters transparency, allowing users to better understand and trust their technology.
4. Decentralization. Data is distributed among users rather than stored on centralized servers, enhancing privacy and security.

At its core, Web 3.0 seeks to deliver a more personalized and efficient online experience by integrating advanced machine learning and artificial intelligence. AI-powered search engines, virtual reality (VR), augmented reality (AR), and sophisticated data analytics work together to provide consumers with tailored content at unprecedented speeds.

An Innovation Epicenter or a Digital Divide?

Web 3.0's potential to revolutionize the Internet raises important questions: Will this era become an inclusive epicenter of innovation, involving and benefiting everyone? Or will it establish complex reputation-based systems of meritocracy, leaving vast portions of humanity in the shadows?

A key feature of Web 3.0 is its disruption of traditional power structures. Internet giants like Google and Meta will no longer monopolize consumer data. Instead, a new wave of innovative startups is redefining the digital landscape:

- Filecoin and IPFS. These projects revolutionize data storage by turning centralized cloud services like Google Drive into decentralized networks where information is distributed among users.
- Ethereum. A blockchain platform enabling decentralized applications (DApps), Ethereum offers alternatives to the centralized services of the Web 2.0 era.
- Polkadot. This network facilitates interoperability among blockchains, enhancing ecosystem functionality and flexibility.
- Brave. A privacy-focused browser that blocks ads and trackers while rewarding users for engaging with advertisements, Brave exemplifies the user-centric ethos of Web 3.0.

Empowering Users Through Decentralization. Web 3.0's defining feature is its shift toward user empowerment. Under this model, individuals gain unprecedented control over their online identities and data:

- Users store their data on personal devices rather than centralized servers, ensuring greater privacy and ownership.
- Digital identities become self-sovereign, enabling secure access to online services without relying on centralized intermediaries like Meta or Google.
- Legal interactions within Web 3.0 occur as equal private law relations, fostering a decentralized, user-driven ecosystem.

Web 3.0 in Education: Transforming Learning for Children. Web 3.0's impact extends beyond commerce and technology; it also holds immense promise for education. Several projects demonstrate how blockchain-based systems can reshape learning for children:

- **BitDegree**. This educational platform leverages blockchain and tokenization to motivate students by rewarding them with virtual assets for their achievements. These tokens can serve various purposes, encouraging engagement and learning.
- **Teachur**. By utilizing blockchain to track achievements and create personalized educational paths, Teachur caters to the diverse interests and needs of the younger generation, ensuring tailored and meaningful learning experiences.

A Vision for a Decentralized Future. Web 3.0 represents the third generation of the Internet, redefining how we interact with digital spaces. By shifting power from corporations to individuals, it creates a more open, secure, and user-centric web. In this new digital landscape, users own their data, their identity, and their online lives.

As the era of Web 3.0 unfolds, its potential to empower humanity will depend on how well its principles are integrated into systems of education, governance, and law. By embracing these changes, we can ensure that the digital world serves as a tool for growth, equity, and innovation, rather than a new frontier of inequality.

Metaverse. A Digital Reality Beyond the Known. The metaverse is often described as a digital analogue of reality, a boundless virtual environment where individuals interact to work, play, conduct business, and socialize. It is a realm where the boundaries of physical and virtual worlds blur, enabling immersive experiences through advanced technologies. As highlighted in the most comprehensive dictionaries of Web 3.0, the metaverse operates with its own internal currency systems, typically based on blockchain technology, to facilitate transactions such as buying, selling, or exchanging goods and services[40].

The term "metaverse" is derived from the prefix "meta," meaning "beyond," and "universe," referring to all that exists in space and matter. True to its name, the metaverse transcends the limitations of our physical world, offering a new dimension of existence.

The Metaverse and Corporate Ambitions

[40]N. Beutin, D. Boran, The great Web 3.0 Glossary. Deutscher Fachverlag GmbH, Fachmedien Recht und Wirtschaft, Frankfurt am Main. 2023. – 199 p.

Large tech companies have embraced the metaverse, fueling its rapid growth and popularity. This digital realm is characterized by several key attributes:

1. Persistence: It exists continuously, without interruptions, irrespective of the user's presence.
2. Real-Time Interaction: Users engage in live, synchronous activities, mimicking real-world interactions.
3. Economic Integration: The metaverse includes complex economic systems with blockchain-based currencies, enabling commerce and ownership.
4. Physical-Virtual Bridging: It connects the physical and digital worlds, merging the tangible with the intangible.
5. Open Content: User-generated and community-driven content defines its ever-evolving nature.

These features make the metaverse a powerful ecosystem where reality is augmented, yet the implications of this augmentation demand scrutiny.

The Depth of Data in the Metaverse. Operating in the metaverse entails unprecedented levels of data collection and processing. A staggering array of personal, behavioral, and physiological data is gathered to power these immersive experiences. The Metaverse collects and processes a huge amount of data about a person, such as: biometrics, facial expressions, eye movements, hand movements, iris movements, speech, brain wave patterns, habits, choice patterns, user activities, behavior patterns, feelings, expressions and body language, user conversations, digital footprint on the Internet, body movements, cultural data, financial data, connections, location, age, purchasing preferences, medical data, digital assets, identity of virtual objects, accounting of cryptocurrency spending, physiological data, physical data[41].

This includes:

- Biometric Data. Facial expressions, eye movements, iris scans, and speech patterns.
- Behavioral Data. User habits, decision patterns, and activity logs.
- Emotional and Physical Insights. Body language, feelings, and physiological states like brain wave patterns.

---

[41]Yavuz Canbay, Anil Utku & Pelin Canbay, 'Privacy Concerns and Measures in Metaverse: A Review' 15th International Conference on Information Security and Cryptography (Conference Paper, 2022) 84.

- Cultural and Financial Data. Purchasing preferences, connections, and cryptocurrency usage.
- Health Information. Medical records and physiological responses.
- Digital Footprints. Internet activity, digital assets, and the identity of virtual objects.

This level of data collection raises profound ethical and legal questions about privacy, consent, and the balance of power between users and corporations.

A Call for Meta-Rights in Meta-Reality. The term "metaverse" signifies going beyond the visible and the known. But as the metaverse pushes these boundaries, it also risks intruding into the most intimate aspects of our lives, creating complex and unpredictably detailed digital profiles. Are our fundamental dignity and rights safeguarded in this digital reality?

This emerging landscape challenges us to envision **meta-rights**: a new category of human rights tailored to protect individuals within these virtual dimensions. Meta-rights would address critical concerns such as data sovereignty, algorithmic fairness, and the ethical use of digital identities. They are essential to shaping a legal **meta-narrative** that ensures the metaverse evolves as a fair and just digital society.

The metaverse is a gateway to transformative opportunities, but it also demands vigilance and thoughtful regulation. By recognizing the need for meta-rights, we can strive to create a digital ecosystem that respects individual dignity, safeguards personal freedoms, and lays the foundation for an equitable digital future.

Industry 4.0. Industry 4.0, often referred to as the Fourth Industrial Revolution, signifies a transformative era characterized by the integration of advanced technologies into manufacturing and industrial practices. Key components of this revolution include the expansion of computing power, rapid advancements in software development, the evolution of human-machine interfaces, progress in artificial intelligence (AI), and the emergence of quantum data processing.

Key Trends in Industry 4.0:

1. Advancements in Computing Power and Software Development. The exponential growth in computing capabilities has facilitated the development of sophisticated software solutions, enabling more efficient and streamlined industrial processes. This includes the

simplification of software interfaces, making technology more accessible across various sectors.

2. Evolution of Human-Machine Interfaces. Innovations in interfaces, such as gesture and thought-based controls, are revolutionizing the way humans interact with machines, enhancing efficiency and user experience in industrial settings.
3. Artificial Intelligence and Algorithmic Progress. Significant strides in AI and algorithm development are leading to smarter systems capable of optimizing operations, predicting maintenance needs, and improving decision-making processes within industries.
4. Quantum Data Processing and Storage. The development of quantum computing and advanced data storage solutions promises to revolutionize data handling capabilities, offering unprecedented speed and security in information processing.

According to the Global Innovation Index 2024[42], technological progress remains robust, particularly in areas like genome sequencing, computing power, and electric batteries. However, there are signs of deceleration in innovation investments, with a notable 5% decrease in scientific publications between 2022 and 2023, marking a deviation from the decade-long average increase of around 4%.

The application of Industry 4.0 technologies holds substantial potential across various sectors, including green technologies, energy, healthcare, and education. In the long term, advancements in these areas are expected to contribute to a healthier population and a cleaner environment, thereby enhancing productivity and supporting the achievement of the United Nations Sustainable Development Goals (SDGs).

Despite these technological advancements, there remains a noticeable lag in the legal frameworks governing human and child rights on a global scale. The rapid pace of innovation necessitates a concurrent evolution in legal regulations to ensure that the benefits of Industry 4.0 are equitably distributed and that the rights of all individuals, especially children, are adequately protected.

In conclusion, while Industry 4.0 presents numerous opportunities for economic and societal advancement, it also poses challenges that require careful consideration. Balancing technological

[42] https://www.wipo.int/web-publications/global-innovation-index-2024/en/global-innovation-tracker.html?utm

progress with the protection of human rights is essential to ensure that this new industrial era leads to inclusive and sustainable development.

The Innovative Wave of Law, Education and Institutional Development as a Framework for Digital Sustainability.

Modern globalization, once rooted in economic and cultural exchanges, has evolved into a predominantly digital phenomenon. This shift marks a new phase in human history, as highlighted by scholars emphasizing the transformative nature of globalization on the state and law. Scholars of globalization and international law have long argued for the importance of interdependence and cooperation in addressing global challenges. Today, these challenges are magnified by the rapid and uneven digitalization of societies across the planet[43].

In his work Novacene[44], James Lovelock explores the emergence of a new era following the Anthropocene, which he defines as the "era of technological mastery of the patterns of organization of matter and energy we call information." This perspective underscores how digital technologies, from artificial intelligence to big data, are reshaping every aspect of human existence.

The Digitalization of Knowledge and Industry

The integration of technology into fields like medicine illustrates this transformation. Previously, drug development was confined to the domain of biologists and chemists. Today, it is a multidisciplinary endeavor involving mathematicians, data scientists, and programmers. This technological expansion is not limited to healthcare but is a prerequisite for success across industries. Companies that rapidly adopt and deploy cutting-edge digital tools have a competitive edge in global markets.

However, technological advancements alone do not guarantee the expansion of human freedoms. Despite the undeniable progress in technological fields, the regulation of digital social relations often lags far behind. The uneven distribution of technological benefits exacerbates global inequalities and increases the risk of technologies being misused. Without equitable frameworks, digital innovation may serve to widen the gap between those who have access to these tools and those who do not.

---

[43]Marchenko M.N. State and Law in the Context of Globalization / M.N. Marchenko. – M. 2009. – P. 282–294.

[44]James Lovelock, Novacene (2019). Lovelock J. (with Appleyard B.). Novacene: The Coming Age of Hyperintelligence. Allen Lane, 2019.

The Voice of the People and the Digital Divide There is a troubling disconnect between technological innovation and public participation. The development, implementation, and use of new technologies are frequently undertaken without meaningful public consultation or a clear understanding of societal needs. This lack of inclusivity is especially concerning in the context of children's rights, as digitalization often imposes limitations rather than enhancing freedoms.

Reports from international bodies and children's rights organizations highlight the challenges children face in the digital era. These include limited access to essential technologies, exposure to online risks, and inadequate protection of their digital identities. Global efforts to address these issues must prioritize children's perspectives, ensuring that their voices are integral to shaping policies and systems that affect their lives.

A Call for Digital Legal Frameworks. The lag in legal regulation reflects a broader issue: the absence of robust global mechanisms to govern the ethical use and equitable distribution of technology. As the digital wave transforms economies and institutions, it is critical to establish legal and educational systems that support digital sustainability. Such frameworks must focus on protecting human dignity, expanding freedoms, and ensuring equitable access to technological opportunities for all, particularly vulnerable groups like children.

In this era of rapid digitalization, the global community must strive not only for technological advancement but also for the creation of inclusive, rights-based frameworks that guide its implementation. By fostering transparency, accountability, and public engagement, we can ensure that the digital revolution benefits humanity, rather than deepening existing divides[45].

Advancing Children's Information Security and Global Legal Frameworks. The *Working Group on Children's Information Security* under the Commissioner for Children's Rights in Russia has identified several pressing areas of concern and opportunities for improvement in safeguarding minors in the digital realm. These include:

Risks in Social Networks. Addressing dangerous groups and harmful content targeting children online.

---

[45]Report on the activities of the Commissioner for Children's Rights under the President of the Russian Federation in 2021 / URLhttp://deti.gov.ru/detigray/upload/documents/August2022/OucV7OrXsDXYb6xBrHFF.pdf(date accessed: 14.12.2024).

Age-Appropriate Content. Advocating clear and enforceable age markings for information products on the Internet.

Charter for Internet Content Security. Developing guidelines to enhance online safety for users.

Data Protection. Strengthening protections for minors' personal data in digital formats.

Biometric Risks. Mitigating dangers associated with the use of biometric identification technologies for minors.

Additionally, there are emerging proposals aimed at enhancing the digital safety of children, such as creating separate social networks for primary and secondary school students and specialized devices with protective features, paired with dedicated tariff plans to mitigate digital threats.

Global Law and the Rights of the Child. The realization of children's rights is intricately tied to the broader development of *global law*, which is still in its theoretical and formative stages. As we transition deeper into the digital age, the complexities of technological innovation demand a cohesive and global approach to legal regulation[46][47].

Most contemporary concepts of global law arose in response to challenges unique to the early 21st century[48], such as the rise of the information economy, increasing interconnectedness, and technological revolutions. Several scholars highlight: 1. The Role of Economic Crisis: Emphasizing the necessity of a global legal framework to address crises and promote justice, particularly in combating crimes against humanity. 2. Global Financial Regulation: Proposals for creating unified global financial institutions and corresponding administrative legal systems.

An emerging vision for global law, described by some legal theorists, highlights the growing insufficiency of standalone national legal systems. Instead, they argue for a model of *interconnected, overlapping institutions, norms, and processes* that transcend state boundaries.

---

[46]Rafael Domingo The New Global Law, 2010, 240 p.

[47]Rafael Domingo "Garzón lo hizo muy bien en el caso Pinochet." 31 March 2010.

[48]Michael S. Barr. "Global Administrative Law and the Post-Crisis Financial Order." Proceedings of the Annual Meeting (American Society of International Law) 108 (2014): 31-33. Accessed June 20, 2021.

Early Research in Global Legal Regulation. In the United States[49][50][51], early research into global law is being undertaken at several leading institutions.

The *Institute for International Law and Justice* focuses on the principles of global administrative law, exploring governance in an increasingly interconnected world.

The *Guarini Institute for Global Legal Studies* delves into issues such as addressing *data inequality* on a global scale and ensuring that digital rights, including those of children, are recognized and enforced internationally.

The Path Forward. The integration of children's rights into global legal frameworks requires proactive steps to harmonize existing national laws with emerging global standards. It is essential to address challenges posed by digitalization while ensuring justice, equity, and the well-being of vulnerable populations such as children. In doing so, global law must not only adapt to technological progress but also actively shape it to safeguard humanity's core values and rights.

Concluding the Paragraph: A Vision of Global Law and Digital Transformation.

Globalization underscores the importance of children's rights and elevates the role of law as a vital regulator of modern globalized social relations in the pursuit of sustainable development. Law, in its most contemporary understanding, is a dynamic system encompassing both international and domestic forms, designed to actualize universal human rights. This framework aims to guide humanity toward sustainable development, resolving global challenges and fostering a more equitable civilization. At its core, the rights of children and humans provide individuals with access to essential social benefits, secured and upheld by this integrative legal system, which forms the bedrock of emerging global law.

The global digital transformation of our time has a profound impact on children's memory, emotions, intelligence, and imagination,

---

[49]Walker N. intimations of global law. - Cambridge: Cambridge univ.. Press, 2015. - 212 p

[50]Institute for International Law and Justice / URL:https://www.iilj.org/(date accessed: 14.12.2024).

[51]Guarini institute for global legal studies / URL:https://www.guariniglobal.org/(date accessed: 14.12.2024).

shaping their capacity to thrive in a rapidly changing world. This makes the study of digital human and child rights particularly relevant within the frameworks of Industry 4.0 and the future Web 3.0.

Imagine the year 2045—a future where technology embodies humanity's brightest aspirations, shaped by deliberate and ethical choices in digital transformation. In this vision, we see a unique confluence of the human and technical worlds, creating a harmonious synergy of abundance through digital sustainability. This scenario ensures the realization of fundamental rights like freedom of assembly and freedom of speech, redefined for a new era of individual and collective freedom.

The rigid vertical structures of old—dominated by narrow, fictional narratives imposed by the few—are replaced by diverse, fluid networks. These "connectives" unite the voices of millions into a vibrant, inclusive social fabric. Boundaries dissolve as individuals from every corner of the planet forge meaningful connections, forming startup cities and networked institutions that transcend geography. Though digital in form, these bonds are undeniably real, driven by shared aspirations and collective progress.

Voices once silenced or tokenized within hollow democratic frameworks now resonate on platforms of techno-legal participation. This is no longer the superficial engagement of the past but an authentic form of interaction where every voice holds significance, every opinion shapes outcomes, and every individual actively contributes to the ***chaosmos* of social evolution**.

The transition from hierarchical verticals to egalitarian horizontals reshapes the landscape of Web 3.0. In this rhizomatic structure, traditional peripheries take center stage at the stakeholder table, creating a dynamic digital ecosystem where individuals of all ages contribute uniquely to global prosperity. This shift emphasizes inclusivity and decentralization, empowering everyone to have a stake in the world's future.

If globalization in the early 21st century required humanity to think and act on a planetary scale, the digital globalization of today demands integration, agility, and the forging of alliances with emerging technological entities. These partnerships enable all individuals to participate in addressing global challenges. With this foundation laid, we now turn to the critical concepts of digital human rights and the digital rights of the child.

1.3. Mapping the Evolving Landscape of Digital Human Rights and the Digital Rights of the Child

The notion of digital human rights, and particularly the digital rights of children, marks a new frontier in legal scholarship. In this study, we treat children's rights in the digital sphere as a vital component of a broader global framework of **digital human rights**[52]. To ground our exploration, we first clarify the key concepts at stake—**digital human rights, digital rights of the child**, and related ideas such as **connectivity, the network state, digital law, startup cities, and sustainable innovation ecosystems**.

In international discourse, the term "digital rights management" (DRM) has long referred to methods of protecting copyrights for digital media through technologies that restrict unauthorized copying and use—an approach primarily recognized in Web 3.0 contexts as outlined in specialized dictionaries[53]. However, DRM remains confined to the private-law domain within the emerging Web 3.0 economy.

By contrast, digital rights in a broader sense rest on the principles of open communication, equality, and technological accessibility (including access to Global Digital Commons). They encompass freedoms of expression, privacy, and information access—values highlighted by various vocabularies of technical reality. A philosophical dictionary further refines these ideas, noting that digital human rights represent "the extension and application of universal human rights to meet the needs of an information-based society." Yet it is crucial to observe that many current definitions often substitute the term "information rights" for "digital rights." [54]. While information and data undeniably drive modern societal transformations, they represent only one dimension of a more profound shift affecting both society and

---

[52]Burianov M.S. Digital human rights in the context of global processes: theory and practice of implementation: monograph / M.S. Burianov - Moscow: RUSAINS, 2022 - 148 p. - P. 120.

[53]N. Beutin, D. Boran, The great Web 3.0 Glossary. Deutscher Fachverlag GmbH, Fachmedien Recht und Wirtschaft, Frankfurt am Main. 2023. – 199 p.

[54]Philosophical Dictionary / – URL:http://slovariki.org/filosofskij-slovar/14089(date accessed: 14.12.2024).

human rights. Indeed, legal, political, and scholarly reference works have yet to provide a fully realized definition of these emerging categories[55]. The work of legal scholar D. A. Savelyev underscores the importance of these concepts by asserting that information and the information rights of the individual, in addition to supporting personal development, education, culture, and communication, serve as guarantees for the exercise of universally recognized rights and freedoms[56]. Subsequent contributions by N. V. Vinogradova and R. G. Vakhrameev further shaped our understanding of information rights and the right to information[57][58].

Notably, I. Yu. Pashchenko points out that digital rights, unlike many information rights, find their primary fulfillment within virtual environments[59]. By contrast, some information rights can be realized in tangible form outside the digital realm, producing similar outcomes. In the digital context, the scope and implementation of these rights often hinge on the frameworks set forth by information system operators, whose user agreements effectively establish the rules of engagement.

Digital human rights thus broaden the horizon of universal human rights, long guaranteed by international and national laws across the globe. V. D. Zorkin, Chairman of the Constitutional Court of the Russian Federation, has argued that digital rights must be acknowledged and operationalized through legal mechanisms, including law enforcement and judicial decisions[60]. Although his perspective

---

[55]Philosophical Dictionary / – URL:http://slovariki.org/filosofskij-slovar/14089(date accessed: 14.12.2024).

[56]Saveliev D.A. Human rights in the field of information: International legal aspects: Diss... Cand. of Law. Sciences – M., 2002. – P. 6.

[57]Vinogradova N.V. Legal mechanism for protecting information rights and freedoms of man and citizen in the Russian Federation: Diss... Cand. of Law. Sciences – Saratov., 2011. – P. 20.

[58]Vakhrameev R.G. The right to information in the Russian Federation (constitutional and legal research): Diss... Cand. of Law. Sciences – Ekaterinburg., 2015. – P. 20.

[59]Pashchenko I.Yu. Information as an object of public-law regulation in the context of digitalization: diss... candidate of legal sciences: Federal State Budgetary Educational Institution of Higher Education "Kuban State University". 2022. – 300 p. – P. 30.

[60]Law in the Digital World – URL:https://rg.ru/2018/05/29/zorkin-zadacha-gosudarstva-priznavat-i-zashchishchat-cifrovye-prava-grazhdan.html(date accessed: 14.12.2024).

encouragingly embraces the expansion of fundamental rights in the digital era, it remains predominantly tied to the conception of information rights derived from the Industry 3.0 paradigm.

It seems timely, then, to define "human information rights" as the capacity for private and secure engagement with the information infrastructure, media, and communications. Such a definition does not undermine universal human rights; rather, it enables their fruition in a digital environment.

Crucially, one of the key priorities of states and international organizations should be to realize and safeguard digital human rights—envisioned here as the opportunities and freedoms that shape our digital identities—rather than focusing merely on cybersecurity. Drawing on A. K. Zharov's dissertation[61], we note that cybersecurity is often defined as the protection of individuals, society, and the state from internal and external threats that undermine the availability, confidentiality, security, and resilience of information infrastructures and the trust-based digital environment. While cybersecurity is essential, it is the effective implementation of digital human rights themselves that can ensure not only the constitutional liberties and well-being of citizens, but also the sustainable social and economic development of nation-states in an increasingly interconnected world.

According to the Civil Code of the Russian Federation, "digital rights" are defined as property rights that exist within an information system. These legally recognized rights, while important, are currently limited in scope[62]. At their core, they enable transactions and the disposal of digital property, but do not yet fully capture the broader social, political, and cultural dimensions of digital existence.

In 2020, the Coconet: Digital Rights Camps project[63] undertook a study in the Asia-Pacific region to explore how local communities define

[61]Zharova A.K. Theoretical foundations of legal regulation of the creation and use of information infrastructure in the Russian Federation: author's abstract. dis. ... Doctor of Law: M. 2020. - 388 p. - P. 340.

[62]Sitdikova, RI, Sitdikov. RB Digital rights as a new type of property rights [Tsifrovyye vid imushchestvennykh prav]. Property relations in the Russian Federation, 9 (in Russian). 2018.

[63]Coconet: What are digital rights? – URL: https://www.apc.org/en/news/coconet-what-are-digital-rights (accessed: 14.12.2024).

digital rights. Participants were asked to identify the key rights they sought, and several insights emerged:

- Digital rights embody the application of universal human rights in digital spaces.
- They represent the right to express oneself safely, privately, securely, and sustainably online.
- They are fundamental and inherent, promoting inclusion, equality, and equitable access to both infrastructure and information.
- Digital rights encompass freedoms of speech, expression, association, and assembly. They cover access to internet-enabled devices, the right to information, and participation in digital platforms.
- Digital rights are relevant to all—both ICT users and non-users—and empower individuals vis-à-vis companies, encouraging equal and fair participation.
- They ensure personal control, autonomy, and agency in digital spaces, and protect against the privatization, monopolization, and monetization of human life online.

Notably, one prevailing approach to digital rights remains relatively narrow, focusing primarily on issues like the right to access the Internet[64] and the right to be forgotten[65], or simply digitizing certain preexisting rights[66].

---

[64]Varlamova, N.V. Digital rights - a new generation of human rights? // Proceedings of the Institute of State and Law of the Russian Academy of Sciences. 2019. No. 4. Electronic resource. - URL: https://cyberleninka.ru/article/n/tsifrovye-prava-novoe-pokolenie-prav-cheloveka (date of access: 12/14/2024).

[65]Talapina, E.V. Evolution of human rights in the digital age // Proceedings of the Institute of State and Law of the Russian Academy of Sciences. 2019. No. 3. Electronic resource. - URL: https://cyberleninka.ru/article/n/evolyutsiya-prav-cheloveka-v-tsifrovuyu-epohu (date of access: 05/25/2023).

[66]Karasev, A.T. Digitalization of legal relations and its impact on the implementation of certain constitutional rights of citizens in the Russian Federation // Antinomies. / A.T. Karasev, O.A. Kozhevnikov, V.A. Meshcheryagina 2019. No. 3. Electronic resource. - URL: https://cyberleninka.ru/article/n/tsifrovizatsiya-pravootnosheniy-i-ee-vliyanie-na-realizatsiyu-otdelnyh-konstitutsionnyh-prav-grazhdan-v-rossiyskoy-federatsii (date of access: 12/14/2024).

However, international developments suggest a broader and more progressive understanding of these rights:

- Several countries have recognized internet access as an inalienable human right. For instance, **Estonia** did so in 2000, followed by **France** in 2009, and **Costa Rica** in 2010[67].
- Since July 1, 2010, **Greece** has constitutionally guaranteed each citizen the right to participate in the information society. By 2013, **Mexico** had amended its constitution to establish access to the information society and the Internet as inalienable human rights.

France offers a particularly forward-looking framework, proposing three "generations" of digital rights[68]:

1. **First Generation:** Ensuring accessible digital communication with government, protecting personal data, guaranteeing cybersecurity, providing access to digital signatures, and supporting citizens with special needs.
2. **Second Generation:** Fostering digital identity, universality, accountability, multichannel access, and transparency.
3. **Third Generation:** Embracing targeted, once-only data usage; active and effective digital services; transparent and informed data use; open algorithms; the governance of artificial intelligence; and the ownership and management of personal data.

What emerges from these discussions is a trajectory of conceptual evolution. Initially, digital rights were primarily understood as property-related or as extensions of established legal frameworks into digital spaces. Over time, however, the notion of digital rights has expanded significantly, becoming a vehicle for ensuring that universal human rights—such as freedom of expression, privacy, and access to information—are fully realized in our increasingly technology-driven world[69].

The General Data Protection Regulation (GDPR), adopted by the European Union in 2016, introduced a robust set of rights related to

---

[67]Saidov A.Kh. Digital Rights As Inalienable Human Rights // Trudi po Intellectualnoy Sobstvennosti (Works on Intellectual Property). 2023. Vol. 44(1). P. 32–39.

[68]OECD. Axer le secteur public sur les données: marche à suivre. 2020 – 196 p.

[69]OCDE (2019[13]), Digital Government Review of Panama: Enhancing the Digital Transformation of the Public Sector, https://doi.org/10.1787/615a4180-en

personal data protection. These rights include access to one's data, the right to rectification, the right to erasure, and others, each intended to empower individuals and enhance trust in the digital environment.

Concurrently, a global debate has emerged over whether we need new, neuro-specific human rights—rights that some countries, such as Spain, consider part of the broader spectrum of digital human rights (as reflected in Spain's Charter of Digital Rights). This conversation underscores a growing consensus on the need to protect and promote human rights in a rapidly changing technological landscape. UNESCO, for example, advocates for a global ethical framework to guide national policies and regulations in safeguarding individual freedoms when it comes to neurotechnologies, particularly in non-medical contexts.

Defining digital human rights, however, remains challenging. The term "digital" is often narrowly associated with the online world or the Internet, yet digitalization extends well beyond these confines. For example, biometric data can exist without an internet connection, and "digital" does not necessarily exclude analog technologies. *Another conceptual hurdle is the tendency to treat the digital and physical worlds as binary opposites—online vs. offline, digital vs. analog or physical, real vs. virtual. As the boundaries blur amid the Fourth Industrial Revolution, it is increasingly difficult to classify any aspect of modern life as strictly real or virtual. Does virtual harassment become any less "real" simply because it occurs online rather than face-to-face?*

The advent of the Internet created a virtual dimension that mirrors the physical world. Humanity now inhabits both the tangible, metric space and an intangible, borderless digital topology. Yet despite people's deep immersion in digital realms, legal norms have not always kept pace, and the Internet remains largely governed by the "law of code" and mathematics rather than clearly articulated legal frameworks. Although the United Nations is contemplating basic human rights in the context of the Internet, we are far from achieving the necessary conceptual clarity, let alone meaningful legal regulation.

Meanwhile, the digital revolution did not end with Industry 3.0; it merely laid the groundwork for Industry 4.0 technologies—machine learning, artificial intelligence, big data, blockchain, virtual and augmented reality (VR/AR), the Internet of Things, biometric systems, drones, neurotechnologies, and beyond. These innovations do not always require an internet connection, further expanding what "digital" can entail. As a result, our understanding of digital human rights must be broader and more inclusive than ever before. If human information rights primarily addressed access to information and the Internet, digital human rights today encompass opportunities enabled by Industry 4.0. These rights face unprecedented challenges and global digital problems, straddling issues of security, privacy, equality of access, and the prioritization and ethical development of emerging technologies.

A pressing question thus arises: **Are digital human rights universal rights that belong to every individual, a groundbreaking new generation of rights, or a transformative era of human rights that subsumes all previous ones**? In this study, I approach **digital human and child rights as heralding a new era**—one defined by the absorption of all facets of human rights into the digital sphere. This era elevates our capacity to address global challenges through advanced technologies such as artificial intelligence, quantum computing, big data, biotechnology, Web 3.0, and the incorporation of biometric, genetic, and medical data into the human rights framework.

Notably, advocates for a decentralized economy often lack a foundational understanding of human rights, raising concerns about the trajectory of Web 3.0. Contrary to the hope that Web 3.0 might rectify the shortcomings of earlier Internet phases, some of its supporters—at least in Russia—risk conflating human rights with other concepts, diluting their significance. They may also prioritize robot rights over human rights or fail to maintain a coherent, systemic approach to ensuring fundamental freedoms. These tendencies emerge from a blend of meritocratic views, post-humanist leanings, and an absence of contemporary legal reasoning. Indeed, few IT and Web 3.0 dictionaries reference human or child rights, revealing a critical gap in the theoretical and legal discourse—a gap this research aims to fill. By doing so, we hope to prevent the negative consequences that could arise during the forthcoming digital transformations.

Our text highlights the complexity of defining "digital" and integrating it into existing human rights frameworks. As technology evolves, so too must our conception of rights. The transformation from Industry 3.0 to 4.0 technologies underscores that digitalization has escaped the narrow confines of the Internet age and now permeates every aspect of our lives—economic, social, political, cultural, and biological.

The call for neuro-specific human rights and the focus on non-medical uses of neurotechnology illustrate how today's moral and legal dilemmas extend beyond data protection into the realm of human cognition and identity. This shift from simple online privacy concerns to deeper ethical questions about the human mind and body signals a new frontier in rights discussions.

Our approach also warns of the dangers inherent in placing blind faith in emerging economic models (like decentralized economies) and new technological paradigms (like Web 3.0) without ensuring they are grounded in robust human rights principles. The future of human rights in the digital era depends on a nuanced and dynamic understanding—one that is flexible enough to adapt to rapid technological change yet anchored in enduring principles that protect human dignity and autonomy.

Challenges in Defining Digital Human Rights[70]**:**

1. **Conceptual Narrowness**. Current interpretations of digital human rights often focus too narrowly on information rights or intellectual property protections. This limited view ignores the broader spectrum of social, political, and cultural dimensions encompassed by digitalization[71].
2. **Private-Law Perspective**. Some definitions frame digital rights primarily within the confines of private law, reducing them to property rights or contractual interests. This approach overlooks the inherently public and global nature of digital ecosystems, where questions of governance, societal welfare, and collective identity become equally important.
3. **Internet-Centric Frameworks**. Equating digital rights solely with human rights "online" or within internet platforms restricts understanding. As we move toward Web 3.0, augmented reality, and metaverses, the digital landscape stretches beyond traditional internet boundaries. The concept of digital rights must adapt to these evolving frontiers rather than remain tied to outdated notions of cyberspace.

Global Digital Identity and Citizenship. The discussion of digital human rights is closely tied to the concept of digital identity. The ease

---

[70]Burianov M.S. Digital human rights as a condition for the effective participation of Russia and other member states of the Eurasian Economic Union in digitalization 4.0 // Technical and technological problems of service. 2021. No. 2 (56). - P. 83-90.

[71]Buryanov, M. S. Gateway to Global Law: Global Digital Human Rights // Scientific Works of the Moscow Humanitarian University. 2020. No. 2. - P. 63-66.

with which minors adapt to digital environments influences evolving notions of digital citizenship—an area where privacy, freedom of expression, and personal agency collide with national jurisdictions and global governance structures[72].

As global citizenship becomes more relevant in a highly interconnected world, we must reconcile the principles of digital citizenship with classical legal frameworks, such as those outlined in the 1933 Montevideo Convention. Modern network-based governance models call for principles like accountability, transparency, and decentralization, reflecting the need to update our public legal sphere in accordance with 21st-century technological realities.

Global Digital Rights and Public Sphere Efficiency. Global digital human and child rights can help us address the large-scale digital challenges confronting our civilization. Initiatives like the GovStack[73] project illustrate how these rights, if properly recognized and implemented, might enhance the efficiency of governance systems and ensure that knowledge—once held by an elite few—no longer serves as a hegemonic tool. Instead, digital rights could enable a more decentralized, equitable relationship between individuals, corporations, and states, preventing the commodification of personal data and protecting users from subtle forms of manipulation and exploitation.

Impact on Digital Legal Relations and AI Governance.

The debate over digital legal relations, especially as it pertains to entities like "robots, digital personalities, and big data operators,"[74] must be grounded in an awareness of digital human and child rights. Similarly, establishing the legal status of artificial intelligence (AI)[75] systems involve far-reaching consequences for how we treat the outputs of these technologies. This includes grappling with ownership and copyright issues related to AI-generated content.

---

[72]Swarts G. Re/coding Global Citizenship: How Information and Communication Technologies have Altered Humanity… and Created New Questions for Global Citizenship Education URL:https://doi.org/10.46303/ressat.05.01.4(date accessed: 31.05.2023).

[73]Govstack / URL: https://www.govstack.global/ (accessed: 31.05.2023).

[74]Minbaleev A.V. Digital legal relations: concept, types, structure, objects // Digital law: textbook / edited by V.V. Blazheev, M.A. Egorova. - M., 2020. - P. 67.

[75]Law, digital technologies and artificial intelligence: collection of articles / ed. E.V. Alferova. - Moscow: INION RAS, 2021. - 267 p. - P. 188.

Copyright and AI-Generated Content. Currently, there are four main approaches to determining who holds the copyright of AI-assisted creations:

A) **Copyright to the AI Creator**. Awarding copyright to the AI's creator assumes that the developer, by building the algorithmic tool and training it on vast datasets, "authored" the result. However, this stance is flawed if data collection violated privacy norms or if the developer keeps the algorithm's principles hidden from scrutiny.

B) **Copyright to the Data Owner**. Since data is the core "fuel" for AI outputs, some argue that whoever owns the dataset should hold the rights. Yet this can be problematic if the data was gathered without user consent or clear purpose, potentially undermining trust and privacy.

C) **Copyright to the AI Itself**. Granting AI an independent legal personality or creative agency is untenable at current technological stages. AI lacks self-awareness, moral responsibility, and the capacity for independent authorship that characterizes human creators.

D) **Copyright to the Prompt-Engineer**. The person who crafts the input prompts that guide the AI's output might claim authorship. While this recognizes the human's role in shaping the AI's creative process, it still does not fully address complexities of consent, data provenance, and the interplay of multiple contributors.

Towards a Fifth, Alternative Approach to AI-Generated Intellectual Property

A more holistic vision of intellectual property (IP) in the age of artificial intelligence suggests that AI algorithms, trained on human-generated data—digital footprints, intellectual creations, and sensitive personal information such as biometric, genetic, and medical data—should not serve merely as profit engines for corporations, venture funds, start-ups, or even state actors. Instead, the beneficiaries of AI-driven intellectual activity should be all of humanity, ensuring that the fruits of technological progress do not exclude, marginalize, or displace people.

The automation of social processes using human-generated data must not lead to mass unemployment, digital bureaucratization (e.g., through the imposition of entirely digital currency and infrastructure), or pervasive surveillance enabled by recognition technologies. By embedding the concept of digital human rights into the digital economy, we create conditions where increased economic efficiency is accompanied by maximized freedoms, abundant resources, and equitable benefits—particularly for vulnerable groups, including children.

A Balanced Approach: The ideal solution would acknowledge the contributions of all parties—AI creators, data owners, and prompt-engineers—while being fully compatible with digital human rights. This means ensuring equitable benefit-sharing and respecting privacy. Special care must be taken to safeguard children's data, ensuring that their emerging digital identities are not exploited, and that their rights—current and future—remain intact.

Viewed through the lens of modern scholarship, these issues underscore a profound tension: the digital revolution promises new freedoms, yet also presents risks of unprecedented scale. Think of the digital domain as a new continent being explored: we need rules that respect individual dignity, ensure collective prosperity, and promote a balance between innovation and ethics.

The insistence on "digital human rights" as a distinct and evolving category signals a paradigm shift in how we conceive rights altogether. These are not static entitlements pinned to pre-digital eras; they are dynamic principles shaped by the interplay of quantum computing, AI, and global connectivity. The world we inhabit today—where children navigate digital spaces before they can read a traditional map—requires that our legal, ethical, and philosophical frameworks grow more agile and inclusive.

Ultimately, the success of digital human rights frameworks will hinge on our ability to integrate insights from law, technology, philosophy, sociology, and economics. We must resist the temptation to reduce digital rights to a narrow set of technical or private-law questions. Instead, we need a holistic vision where data governance, neurotechnologies, AI, and emerging metaverses serve the collective good—fortifying individual autonomy, cultural richness, and the ethical use of innovation for generations to come.

Digital Rights of the Child: Moving Beyond Restrictions

Until now, discussions on children's digital rights have too often focused on limitations and prohibitions, framing childhood primarily in terms of what must be safeguarded rather than what can be empowered. In reality, the rights enshrined in the Convention on the Rights of the Child naturally extend into the digital domain. This goes beyond merely protecting children online; it involves recognizing that they have an equal claim to the full range of human rights in digital spaces.

It is imperative to secure a child's right to protection from online risks—exposure to inappropriate content, commercial exploitation, abuse through advertising and marketing, sexual exploitation, pedophilia, and human trafficking. The imperative to prevent violence and discrimination applies as much to virtual settings as it does to physical ones.

Children's Perspectives on AI and the Metaverse.

Today's digital human rights debate intertwines with emergent technologies like artificial intelligence and the metaverse. A survey by the Dutch organization KidsRights[76] reveals that children's priorities revolve around the human qualities they seek in these technologies:

1. **Human Literacy**. Children recognize a lack of inherent "humanity" in robots.
2. **Emotional Intelligence**. They note the absence of genuine emotional capacity in AI.
3. **Love and Kindness**. AI and robots cannot currently exhibit love or compassion.
4. **Authenticity**. Machines lack independent minds of their own.
5. **Human Care and Protection**: Robots cannot truly comfort or console.
6. **Autonomy**. Children insist that robots should not "take over the world."
7. **AI at Work**. Robots should assist rather than replace or marginalize human workers.
8. **Abundance.** Robots should add value to leisure, play, and everyday life.

These eight standards reflect children's hopes and fears. They should inform the framework of children's digital rights in AI-governed spaces, guiding the responsible and ethical design of digital technologies. Crucially, children's opinions must influence how we introduce or withhold certain aspects of their rights from digital conversion, ensuring their developmental needs remain paramount.

For example, certain educational, social, or medical services may be best delivered face-to-face rather than through digital interfaces. The

---

[76]La Fors, K. (2023) 2022 AI Register of Children in The Netherlands: Mapping children's awareness, ethical and social sense-making and imaginaries of artificially intelligent systems via meaningful participation Kidsrights / DesignLab University of Twente; The Netherlands ISBN 9789090371450 (e-book)

objective is not to stifle technological innovation, but to ensure that we maintain meaningful human interaction and do not inadvertently erode vital social functions.

Global Perspectives and Inclusive Debates

Debates over children's digital rights cannot be limited to technologically advanced nations. In the Commonwealth of Independent States (CIS) and other regions with pronounced inequalities, the stakes may be even higher. The World Economic Forum's "Artificial Intelligence for Children Toolkit" (March 2022)[77] echoes this sentiment, highlighting that children in less economically and digitally developed contexts might struggle even more with understanding or exercising their digital rights.

It is therefore essential to shape a global discourse on children's digital rights that emphasizes both protection and empowerment. Rather than focusing solely on cybersecurity and privacy, the principles guiding children's digital rights should also champion the enormous opportunities of Industry 4.0—opportunities that can improve education, healthcare, entertainment, and civic engagement. The design of digital products and services should naturally incorporate these principles from the outset, creating a more equitable and child-centered digital future.

When children articulate their fears and desires regarding AI, they effectively remind us of the primal human need for empathy, authenticity, and care—a reminder that advanced technology is not an end in itself but a tool that must serve human well-being. The argument that all of humanity should be the ultimate beneficiary of AI-generated value underscores a philosophical stance: that modern technological progress, rather than compounding inequality, should level it. The same reasoning must hold true when ensuring children's rights in digital spaces—unless we want a future where technological sophistication flourishes in moral or social desert.

Adopting children's digital rights as a core design principle in technology development promotes innovation that respects human dignity. Rather than reducing young people's online presence to a catalog of prohibitions, we can empower their digital citizenship. We can equip them

---

[77] Artificial Intelligence for Children Toolkit March 2022. World Economic Forum // URL: https://www3.weforum.org/docs/WEF_Artificial_Intelligence_for_Children_2022.pdf (accessed: 17.05.2023).

to navigate, shape, and improve the digital world, recognizing them as tomorrow's leaders, citizens, and creators. Bridging these ethical considerations with international legal frameworks and global dialogues ensures a balanced and inclusive approach—one that marries technological sophistication with timeless human values.

Digital and Decentralized Identity in the Web 3.0 Era

Digital Identity. Digital identity refers to the set of attributes and credentials that represent an individual online. These can range from usernames and passwords to biometric data and other unique identifiers, effectively compiling a "digital footprint" that, in principle, might encapsulate all aspects of an individual's online presence. When we consider children's digital identities, it becomes crucial to minimize data collection—possibly extending protective measures until adulthood is fully reached (for instance, up to age 26)—especially in contexts where reputation-based digital systems may influence future opportunities.

Decentralized Identity (Self-Sovereign Identity, SSI) Web 3.0 heralds a shift toward decentralized identity, empowering individuals to control their own data without depending on central authorities. In a self-sovereign identity model, users retain ownership over their personal information, sharing it selectively and verifying claims via secure, distributed technologies (e.g., blockchain). This shift challenges the historical norm where institutions largely controlled data, enabling a future in which individuals set the terms of data exchange.

Law, Code, and Digital Human Rights

Digital Human and Child Rights. Legal Foundations. Digital human and child rights typically derive from national and international legal instruments. They stipulate basic protections and entitlements for individuals online, safeguarding freedoms and ensuring equitable access to digital tools and services.

Code as Law. Technological Foundations. In contrast, digital identity frameworks—particularly those in Web 3.0—are often embedded directly into technology stacks and network protocols. In these ecosystems, system architecture and code define the scope of rights and opportunities, rather than external laws or regulations. This dynamic can create tension between the "law that defines code" (legislation) and the "code that acts as law" (technological design), necessitating new integrated approaches.

Toward Techno-Legal Platforms and Protocols. To harmonize legal rights with emerging technological capabilities, we need hybrid frameworks—techno-legal platforms—blending regulatory oversight and decentralized system design. Key steps could include:

1. **Hybrid Regulatory Structures**: Develop new forms of regulation combining central and decentralized elements. Blockchain-based approaches can yield transparent and accountable systems that adapt quickly to changing realities.
2. **Common Standards and Protocols**: Establish unified standards for digital identity, interoperability, and other key aspects of Web 3.0 to ensure that different platforms and systems can communicate and align with fundamental human rights.
3. **Liability Mechanisms and Smart Contracts**: Use smart contracts to automatically enforce certain digital rights conditions. For instance, code-based triggers can ensure compliance with human and child rights standards—expanding access and opportunities without compromising legal protections.
4. **Cautious Introduction of Reputation Systems**: Delay the implementation of widespread reputation-based systems until at least 2030. This gives time to build accessible principles and ensure that such mechanisms do not inadvertently erode human rights or exclude vulnerable groups. Coupled with concepts like unconditional basic income, this cautious approach can prevent digital reputations from becoming tools of exclusion in a new, decentralized internet era.

Digitalization of the Public Sector and GovTech.

GovTech and the Digital State. The public sector's embrace of digital technologies—often termed GovTech—encompasses everything from "smart" city infrastructures and data-driven policymaking to AI-driven social services. The UK's evolving definitions of GovTech[78] and its high global ranking for open, accountable data underscore the rapidly changing nature of this field.

[78]What's the definition of GovTech – and how is it changing government // URL:https://apolitical.co/whats-the-definition-of-govtech/(date accessed: 17.05.2023).

GovTech harnesses big data, AI for cost reduction and automation, machine learning for unified services, and chatbots or digital interfaces for accessible citizen communication. Similar innovation movements—Fintech, Wealthtech, Lendtech, Insurtech, Regtech, and LegalTech—illustrate the sweeping influence of technology in traditionally regulated sectors.

GovTech vs. Civic Tech. While GovTech focuses on optimizing government functions and service delivery, Civic Tech empowers citizens to participate more actively in governance. However, leading GovTech definitions rarely mention human rights[79][80]. The absence of explicit human rights language in these frameworks signals a critical need to integrate human and child rights principles more directly into the design, implementation, and evaluation of digital public services.

We stand at a historical crossroads. On one hand, the emergence of decentralized identities and self-sovereign frameworks offers a vision of personal empowerment: individuals could control their data and escape the clutches of centralized authorities. On the other hand, without careful legal-technical collaboration, these very systems could become opaque, unaccountable, and ripe for exploitation.

The introduction of techno-legal platforms challenges the old dichotomy between "law as something external" and "code as something innate to the system." Instead, we need models where law and code co-evolve, ensuring that digital innovation respects human rights, fosters equality, and strengthens democratic values. Achieving this balance requires proactive engagement: establishing norms, standards, and protocols before problems manifest, and building flexibility into our systems so they can adapt as technology evolves.

In this context, GovTech and Civic Tech have tremendous potential. GovTech can streamline public administration and improve service delivery, but if not anchored in rights-based frameworks, it risks becoming a technocratic apparatus that overlooks the human core of governance. Civic Tech, meanwhile, empowers citizens and fosters participatory democracy, yet it too must consider the ethical implications of data-driven decisions, algorithmic oversight, and digital identity systems.

---

[79]Public. Transforming public services with new technologies // URL: https://www.public.io/ (accessed: 17.05.2023).

[80]Govtech research. GovTech: An Emerging Sector Revolutionising Public Services // URL: https://www.govtechresearch.com/ (accessed: 17.05.2023).

Most importantly, these discussions must center on the most vulnerable—particularly children. As digital citizens in training, children are forging digital identities from an early age, often without comprehensive safeguards or recourse. Ensuring that digital human rights encompass children's rights, and that these rights shape the code of emerging platforms, is essential. By doing so, we prevent digital ecosystems from becoming dehumanized domains and instead ensure they remain spaces where empathy, fairness, and human flourishing prevail.

In summary, the integration of legal principles into emerging technologies is not merely a technical challenge: it is an ethical and societal imperative. When done thoughtfully, it can yield a digital world that celebrates innovation and efficiency without sacrificing equity, dignity, and the richness of human experience.

Decentralized Autonomous Organizations (DAOs) and the Future of Governance

The Emergence of DAOs. Decentralized Autonomous Organizations (DAOs) represent a groundbreaking approach to automating and decentralizing organizational structures. By harnessing open-source protocols, smart contracts, and blockchain technology, DAOs can facilitate collaboration, financial transactions, and decision-making without traditional intermediaries. In essence, a DAO is a set of algorithmic rules that allows individuals to interact according to transparent, self-enforcing protocols. These features include:

1. **Reduced Transaction Costs**: The use of blockchain and smart contracts minimizes bureaucratic overhead, streamlining administrative processes and cutting operational expenses.
2. **Organizational Transparency**: As open-source code governs a DAO, any stakeholder can scrutinize the rules and the flow of resources, promoting trust and accountability.
3. **Consensus Mechanisms and Stakeholder Involvement**: DAOs use consensus protocols to align the interests of all participants. Every token-holder can influence policy and governance decisions, fostering equitable participation.

The DAO concept, when combined with digital human rights, could shape future governance models at various scales, including the state level. Some legal scholars, for example in Austria, have begun to consider DAOs as forms of civil partnerships, illustrating the interplay between emerging technologies and traditional legal concepts[81].

Decentralization, Digital Identity, and Human Rights

Decentralized Technologies as a Vehicle for Autonomy Decentralized infrastructures—blockchain, cryptography, and token-based governance—offer robust security and transparency. Individuals gain unprecedented control over their digital assets and identities. This echoes the arguments in The Sovereign Individual (1999), where the authors Davidson and Rees-Mogg envisioned technology as a force transferring power from centralized authorities to individuals. In a decentralized environment, each person's token (a digital political right) enables direct participation in organizational governance through community proposals and voting mechanisms.

The Network State and Digital Transformation Proponents of the "network state" concept envision a new form of social structure defined by shared moral innovation, a distinct sense of identity, virtual capitals, blockchain-based censuses, and integrated cryptocurrencies. Such a state relies on "smart contracts" and decentralized principles rather than the hierarchical models of the 20th century. While 20th-century industrial revolutions enhanced living standards economically, the 21st century demands that technology also transform politics, improving openness, accountability, and global cooperation. Failure to do so may jeopardize future progress and stability.

Digital Human Rights in the Global Digital Compact. The Global Digital Compact adopted in 2024 does not specifically define digital human rights, enumerate digital human rights, or create a robust system for their realization.

Human Rights in the Digital Ecosystem Digital human rights are crucial in the context of the UN's Global Digital Compact and must apply across the digital ecosystem. The Australian Human Rights Commission's 2023 guidance, Human Rights in the Digital Age: Further Submissions to the UN Global Digital Compact[82], highlights

---

[81]Network State // URL:https://thenetworkstate.com/the-network-state-in-one-thousand-words (date accessed: 17.05.2023).

[82]Australian Commission. Human Rights in the Digital Age: Additional Material Submitted to the UN Global Digital Compact, 30 April 2023 // URL:https://humanrights.gov.au/sites/default/files/human_rights_in_the_digital

the need for technology-neutral legal norms. States should ensure data and personal information protection without making individuals bear the primary responsibility for defending their own privacy.

However, Australian experts recommend applying existing human rights frameworks online rather than creating entirely new rights. They argue that rights should be "technology-neutral" and apply broadly, regardless of the medium.

Expanding the Horizon of Human Rights. While the Australian perspective is pragmatic, this view may be too conservative. The right to global Internet access is not simply an extension of the right to information. It is a qualitatively new form of engagement with the world—an environment of global connectivity, immediate interaction, and decentralized knowledge exchange. Similarly, the right to open-source technology can be considered a new human right, reflecting the era of digital freedom and decentralized governance.

As technological innovation accelerates, human rights frameworks must evolve to accommodate novel forms of identity, individuality, and governance. We must consider new rights as we shift from traditional sovereign states toward models of global, decentralized governance—where the "sovereign individual" claims rights previously unimaginable.

Envisioning DAOs and network states as the next frontier in governance underscores a profound philosophical and legal transition. Historically, people formed hierarchical organizations—empires, nation-states, corporations—to coordinate action and pool resources. Today, distributed ledgers, cryptographic security, and open protocols offer pathways to non-hierarchical systems that may ultimately place citizens, not bureaucracies, at the center of decision-making.

The introduction of digital human rights into this domain is crucial. If digital rights remain vague or are treated as mere extensions of existing rights, we risk failing to capture their transformative potential. True digital human rights acknowledge that the digital sphere is not just another platform, but a world-changing environment that shapes education, health, political participation, employment opportunities, and even cultural identity from a very young age.

By contrast, recognizing entirely new rights—such as a right to global Internet access or to open-source technologies—reflects the

_age_additional_material_submitted_to_the_un_global_digital_compact_1_1_1.pdf(date accessed: 17.07.2023).

understanding that today's digital ecosystem is as fundamental to human development as literacy or freedom of expression was in previous eras. The challenge lies in embedding these new rights into techno-legal frameworks that prevent abuses, protect the vulnerable (like children), and ensure equitable distribution of the benefits of these technologies.

The tension between using existing human rights frameworks and forging new digital rights also mirrors a broader societal debate: Should we adapt old principles to new contexts, or forge entirely new principles for unprecedented conditions? Given the pace and scale of digital transformation, it may be wise to do both. We must preserve the continuity of enduring human values while also creating new legal and ethical categories that reflect the distinct nature of decentralized technologies, global connectivity, and algorithmic governance.

In summary, DAOs, network states, and digital human rights are not abstract hypotheticals. They are emerging realities that challenge us to think beyond traditional definitions of sovereignty, identity, and law. The sooner we articulate and codify the rights and responsibilities inherent in these novel systems, the better equipped we will be to guide technological progress toward a future that expands human freedom, fosters collective prosperity, and upholds the fundamental dignity of every individual—starting at birth and extending into the digital realms where we now increasingly live, learn, and thrive.

Conclusion: Defining Global Digital Human Rights and Digital Rights of the Child.

At the dawn of the Fourth Industrial Revolution and in an era of intensified digital globalization, we must recognize that technology is not merely an instrument of convenience or efficiency—it is a catalyst reshaping the very fabric of human society. **Global digital human rights** represent a new generation of rights, guaranteed from birth and anchored in international and national law. They stand at the intersection of technology and fundamental freedoms, extending beyond the concepts inherited from the Third Industrial Revolution, which focused primarily on information rights linked to the Internet and computers. In this new paradigm, the aim is not only to ensure global connectivity and equitable access to emerging technologies, but also to protect individuals from new forms of digital risk, including the erosion of privacy, the abuse of personal data, the rise of digital inequality, and the specter of digital autocracy[83].

The digital rights of the child build upon these principles, recognizing that children inhabit a digital environment from the earliest stages of life. These rights ensure their capacity to survive, develop, and freely express themselves in a technological landscape that simultaneously opens unprecedented opportunities for learning and growth, while posing new threats in the form of predatory commercial practices, exposure to inappropriate content, or insidious surveillance. The advent of Web 3.0 and decentralized architectures like DAOs (Decentralized Autonomous Organizations) may eventually reshape the very notion of citizenship, governance, and political participation, granting children (as future adults) the autonomy to influence techno-legal infrastructures that govern their daily lives.

By acknowledging both global digital human rights and the digital rights of the child, we lay the groundwork for a forward-looking framework of **"techno-legal" platforms**—an evolution in governance that merges legal principles with the flexibility and transparency of decentralized technology. These frameworks should strive not merely to prevent harm, but also to foster human flourishing, fairness, and inclusion. In a globalized digital environment, every step toward greater autonomy and equity must be matched by robust safeguards that prevent the concentration of power and the erosion of trust. If done correctly, the digital environment will not be defined by a "measure of control," but by a new measure of freedom, enriching human experience and strengthening the social bonds that sustain us.

As we move forward, integrating global digital human rights and digital rights of the child into policy and practice will help ensure that the technological transformations of the 21st century do not lead us toward fragmentation or oppression. Instead, these rights can guide us toward a future of meaningful connectivity, sustainable development, and global cooperation—making full use of the digital sphere as a realm of empowerment, dignity, and shared prosperity.

---

[83] Burianov M.S. The Role of the UN in the Formation of a New World Order in the Context of Digital Globalization: From the Law of Might to the Force of Law / Collection of articles by finalists of the A. A. Gromyko CIS Young International Relations Competition 2020 / edited by V. V. Sutyrin, A. S. Peshenkov. Moscow: Institute of Europe, Russian Academy of Sciences, Assoc. Foreign Policy Research. A. A. Gromyko – 2021 – 625 p. P. 71-83.

Key Digital Threats:

1. **Digital Surveillance**: The non-transparent and often extralegal collection of personal data—including biometric and medical data—that infringes on fundamental rights and undermines trust in digital systems.
2. **Digital Militarization**: The deployment of military-grade technologies in digital environments. Autonomous weapons, cyber-attacks, and the deliberate weaponization of networks threaten global security, undermine the neutrality of the technological sphere, and potentially endanger human rights, including those of children.
3. **Digital Divide**: The unequal distribution of digital resources and technologies, impeding access to education, healthcare, and other vital services. This disparity disproportionately affects vulnerable groups and hinders societal development.

Overcoming these challenges requires a robust framework of **digital law** that governs the hybrid social relations emerging in digital contexts. Such a legal architecture must protect both individual and child digital rights, ensuring that technology development aligns with the interests of society at large.

Essential Concepts for the Digital Age

- **Digital Sovereignty**: The capacity for citizens and organizations to exercise full control over their digital data, infrastructures, and interactions. By prioritizing digital sovereignty, societies can safeguard and implement digital human rights more effectively, cultivating sustainable digital communities.
- **Connectives**: Digital societies founded on networked interactions, encompassing virtual communities, social networks, and digital platforms. These environments foster democratic engagement, cultural exchange, and innovation, offering fertile ground for the exercise of digital rights.
- **Startup City**: An innovative model of urban organization rooted in digital transformation and sustainable development. Such cities harness technology to reduce digital inequality, enhance residents' quality of life, and protect their rights.
- **Network State**: A form of governance grounded in decentralization, digital sovereignty, and transparency. Network states leverage technology to deliver proactive digital services and strengthen citizens' rights, ensuring that power dynamics

shift away from top-down hierarchies toward more inclusive and participatory models.

- **Sustainable Innovative Digital Ecosystem**: A holistic assembly of people, digital agencies, technologies, processes, and legal frameworks working collaboratively to create, utilize, and distribute innovative digital products and services. This ecosystem, underpinned by technical and techno-legal standards (as recommended by the International Telecommunication Union, a UN agency), emphasizes compatibility, security, efficiency, and above all, the respect for human rights. It transforms information and technological resources into tangible benefits while maintaining ethical standards and compliance with international law.

In a world where digital and physical spaces increasingly converge, establishing a sustainable, rights-centered innovative digital ecosystem emerges as a vital component of global, equitable development.

Conclusion of Chapter 1.

This chapter has asserted **that digital human and child rights** must be effectively integrated into the development of new technologies through their recognition as a **techno-legal standard**. Embedding digital rights into technical specifications and guidelines ensures their consistent, systematic application throughout the entire technological life cycle.

The UN role in formulating technical standards is central to ensuring interoperability, security, and equality of access. By explicitly including digital rights in these standards, we transform them from abstract ideals into concrete obligations—criteria that must be met at every stage of design, implementation, and operation. This approach shifts the conversation from mere prevention of rights violations to the proactive promotion of human freedoms in the digital age.

This Box 3 Key Definitions and Generations of Digital Rights presents the main block of concepts that we propose to use to understand human rights in the context of digital technologies.

Box 4. Key Definitions and Generations of Digital Rights

Global Digital Human Rights – essential capacities for human development, guaranteed from birth and anchored in international and national law; their realization enables access to social goods through the use of digital technologies and extends fundamental freedoms into the digital sphere.

Digital Rights of the Child – fundamental capacities for survival, development and free expression of persons under 18 in digital environments; grounded in law, attentive to how technology shapes identity and well-being; aligned with the principle of intergenerational equity (UN, 2022 Our Common Agenda).

Digital Law – a system of principles, norms, cryptographic solutions and algorithms governing hybrid social relations to realize digital human and child rights and support sustainable development.

Connectives – digital societies formed by networks of people and organizations (virtual communities, Web 3.0, DAOs, platforms) that enable democratic participation, cultural exchange and innovation across borders.

Startup City – an urban innovation ecosystem (start-ups, accelerators, investors, universities and citizens) designed to reduce digital inequality, prevent surveillance and militarization, and protect rights while expanding access to education, health and opportunity.

Digital Sovereignty – full autonomy and control over one's digital data, assets and infrastructure for individuals, organizations and states, supported by legal, technological (RegTech) and organizational measures.

Network State – a governance model based on decentralisation, personal and digital sovereignty and protection of rights, enabled by blockchain/metaverse technologies, delivering transparency, access to information and rights safeguards.

Digital Social Contract (Metaverse) – a shared agreement across virtual worlds that regulates interactions between digital actors based on rights, freedoms and sovereignty of every participant, including children.

Sustainable Innovative Digital Ecosystem – a human-centered assembly of people, technologies, processes and legal frameworks that turn technological opportunities into tangible public benefits while respecting rights, equity and international law.

Key Digital Threats

• **Digital Surveillance** – opaque, extralegal collection and processing of personal data (including biometric, medical and genetic) via AI-enabled monitoring that undermines privacy, movement, expression, assembly and equality.

• **Digital Militarization** – state or defense deployment of military-grade digital technologies (autonomous weapons, cyberattacks, AI-enabled targeting), creating existential risks and violating the duty to protect rights to life, safety and education.

• **Digital Inequality** – unequal distribution of digital infrastructure, resources and skills, producing disparities in access to education, health, employment and participation; disproportionately affects vulnerable groups, including children.

Generations of Digital Rights (summary)

• **First Generation – Accessibility:** universal connectivity and infrastructure; privacy and data protection; digital literacy and equality; protection from discrimination; digital detox; digital inheritance.

• **Second Generation – Proactivity:** exercise of civil, social, economic, political and cultural rights via technology; equal access to new technological opportunities; digital participation in governance; universal basic income in the age of automation; digital identity and personal sovereignty; quality digital education.

• **Third Generation – Unity:** peace and demilitarization of the digital sphere; participation in global digital governance; environmental sustainability of digital technologies; human-centered, responsible AI; global cooperation on digital rights; inclusive start-up development and self-realization.

Historically, struggles for civil rights—against segregation, apartheid, and the suppression of political freedoms—showed that human rights are not luxuries but necessities. Past generations fought for equality, justice, and freedom because these principles shaped everyday life and future prospects.

Today, as digital and physical worlds converge and AI reaches every sector, vigilance is essential. The promise of digital abundance must not eclipse values won over centuries. By recognizing, codifying, and implementing digital rights—including those of children—we lay the groundwork for a just and inclusive digital future.

In the next chapter, we examine the legal enshrinement of children's digital rights to ensure the next generation inherits universal freedoms, dignity, and opportunity in an ever-expanding digital world.

CHAPTER 2. Building a Global Framework: The Legal Foundations of Digital Rights for Humans and Children

2.1. International Legal Recognition and Consolidation of Digital Rights

Connectivity remains sharply uneven: in 2024 an estimated 5.5 billion people (68%) were online, leaving 2.6 billion offline. For children and young people, the gap is severe: only 33% have internet access at home worldwide, just 6% in low-income countries, versus 87% in high-income ones.

The UN Human Rights Council has repeatedly affirmed that the same rights people have offline must also be protected online (2012, reaffirmed in 2016). In 2011, the Special Rapporteur on freedom of opinion and expression underscored the Internet's transformative role and urged States to make access widely available as a means to realize human rights. (The mandate itself was renewed by HRC resolution 7/36 in March 2008).

Recognizing, codifying and implementing digital rights—including children's rights, is therefore not optional; it is foundational to an inclusive digital future consistent with international human-rights standards.

Technological disparities manifest differently depending on where one lives. In regions embracing rapid digital transformation, children and youth face unique vulnerabilities, ranging from unregulated exposure to online risks to the absence of balanced digital education. Conversely, in areas completely disconnected from the digital world, a lack of technology access stymies economic and social opportunities. And yet, technology does not stand still: the Internet era is now evolving toward a new wave of progress built on artificial intelligence, augmented reality, and decentralized ledgers—each of which demands careful legal regulation to ensure barrier-free, non-discriminatory access regardless of age or developmental stage.

In seeking digital sustainability, we must acknowledge that legal norms play a pivotal role in establishing a shared vision of a new social contract. The United Nations, through foundational international documents, has laid the groundwork for both human rights and the rights of the child—documents that remain essential to any discussion of digital inclusion.

- The UN Charter (1945) sets forth humanity's paramount goal: to reaffirm faith in fundamental human rights, the dignity and worth of each individual, and equal rights for men, women, and nations both large and small.
- The Universal Declaration of Human Rights (UDHR, 1948) crystallizes these universal principles, stating in its preamble that "everyone is entitled to all the rights and freedoms set forth in this Declaration," irrespective of race, color, sex, language, religion, nationality, or social origin. Of special relevance to our digital age is Article 27, which proclaims "everyone has the right freely to participate in the cultural life of the community, to enjoy the arts and to share in scientific advancement and its benefits." Although written in a pre-Internet era, this article foreshadows today's need for equitable digital access and technological development that benefits all.
- Children's rights, in particular, are underscored by the UDHR and were further articulated in the Declaration of the Rights of the Child (1959). Both documents stress that children require special care and assistance, recognizing their future role as independent, capable members of society.

One of the principal architects of the UDHR, Nobel Peace Prize and UN Human Rights Prize laureate René Cassin, characterized the 1948 Declaration as "a document of a new type, which organized humanity created at a time when man's power over nature, thanks to science, had grown enormously and it was necessary to decide what good could be done with it." In essence, Cassin underscored the moral imperative: scientific progress—yesterday's leaps in industrial technology and today's digital transformations—must serve humanity's collective benefit rather than endanger or marginalize certain groups.

Recognizing that more advanced technologies now shape daily life, it becomes vital to integrate children's needs into evolving digital frameworks. The discussions in this chapter will explore how international legal standards can be adapted and expanded to protect digital rights in the twenty-first century. Specifically, we will examine how emerging technologies—artificial intelligence, decentralized platforms, and augmented or virtual realities—intersect with foundational human rights principles, ensuring that no one, especially no child, is left behind in our rapidly accelerating digital era.

It should be noted that the Declaration of the Rights of the Child (1959) plays a vital role in establishing the legal status of the child, clearly stating that the rights of the child must be recognized for all children without any exceptions or distinctions based on race, skin color, sex, language, religion, political or other opinions, national or social origin, property, birth, or any other status concerning the child or their family. **Principle 8** states that "the child shall in all circumstances be among the first to receive protection and assistance." We agree that this principle is critically important in the context of the Internet environment, ensuring that children and their interests are considered in the creation of online content and applications. Additionally, **Principle 9**, which states that "the child shall be protected from all forms of neglect, cruelty and exploitation," should be specifically applied and adapted to digital realities.

Article 15 of the International Covenant on Economic, Social and Cultural Rights (1966) specifies that States Parties recognize the right of everyone to participate in cultural life and to enjoy the benefits of scientific progress, including its practical application. Digital transformation is a result of scientific progress, and many digital products, although aimed at adults, are also used by children. This highlights the need to ensure that digital technologies are safe and accessible for young users.

The right to security is enshrined in **Article 17** of the International Covenant on Civil and Political Rights (1966): "No one shall be subjected to arbitrary or unlawful interference with his privacy, family, home or correspondence, nor to unlawful attacks on his honor and reputation." This emphasizes that every person has the right to legal protection from interference or attacks. This right is crucial for updating state obligations and developing modern implementation methods in the context of Industry 4.0 technologies and big data.

The Declaration on the Right to Development (1986) states that the right to development is an inalienable human right "by virtue of which every human being and all peoples are entitled to participate in, contribute to, and enjoy economic, social, cultural and political development, in which all human rights and fundamental freedoms can be fully realized." This Declaration positions humans as the central subjects of the development process, asserting that individuals must be active participants and beneficiaries of the right to development. Implementing the right to development is especially important in the context of the Fourth Industrial Revolution, which is accompanied by economic growth and socio-political challenges that can particularly impact vulnerable groups such as children.

The Convention on the Rights of the Child (1989) provides fundamental guarantees for the realization of children's rights and their healthy development. In the context of digital rights, particularly regarding information access, **Article 17** is noteworthy: "States Parties recognize the important role of the mass media and shall ensure that the child has access to information and material from a diversity of national and international sources, especially those aimed at the promotion of the child's social, spiritual and moral well-being and healthy physical and mental development."

In summary, international documents emphasize the necessity of protecting children's rights in the digital age by ensuring their safety, access to information, and participation in digital development. These principles form the foundation for creating a safe and supportive online environment where children can learn, grow, and interact securely.

Convention on the Rights of the Child (CRC) and General Comment No. 25 (2021).

Among the most important international legal instruments pertaining to children's digital rights is the **Convention on the Rights of the Child**, alongside its **General Comment No. 25 (2021)** on children's rights in the digital environment. According to paragraph 3 of this General Comment, the digital environment is becoming integral to nearly every facet of children's lives—including, crucially, during times of crisis—since education, public services, and commercial activity increasingly depend on digital technologies.

Paragraph 4 stresses that the rights of every child must be respected, protected, and fulfilled in the digital environment. While meaningful access to digital technologies can facilitate the realization of children's civil, political, cultural, economic, and social rights, failing to ensure such access risks exacerbating inequalities and creating new forms of exclusion.

In essence, **General Comment No. 25** (2021) enjoins States to recognize how innovations in digital technologies can significantly impact a child's life and rights, even when the child in question lacks direct internet access. It also reminds governments and other stakeholders to shape the digital environment in ways that prioritize children's best interests.

Key Principles: Equality, Non-Discrimination, and the Best Interests of the Child

1. **Equality and Non-Discrimination**
   States must ensure that **all children** enjoy equal and meaningful access to the digital environment. Realizing this principle may require public investments—such as creating free, safe access points and promoting affordable technology—to overcome financial, infrastructural, and social barriers.
2. **Best Interests of the Child**
   Because the digital environment was not initially designed with children in mind, assessing the "best interests of the child" is an evolving and context-specific process. States should carefully balance children's rights to participation, expression, and development with the need to protect them from digital harms (e.g., exploitation, data abuse).

Core Digital Rights of the Child.

Based on **General Comment No. 25** and wider human rights discourse, the following digital rights emerge as particularly vital:

1. The Right to Non-Discrimination
   Every child should have equal, effective, and meaningful access to the digital sphere. This implies targeted efforts to remove technical, economic, or policy barriers preventing children—especially those in vulnerable situations—from benefiting safely from digital technologies.
2. The Right to Life, Survival, and Development
   Digital tools can be crucial during crises—natural disasters, armed conflict, or epidemics—by aiding healthcare delivery, providing educational continuity, and facilitating rapid communication. Special attention should be paid to children in their early years, when brain plasticity is highest, to ensure technology use bolsters cognitive, emotional, and social development rather than hindering it.
3. The Right to Be Heard (Child's Opinion)
   When crafting legislation, policies, or programs related to the digital environment, States must actively involve children, valuing their insights on usability, safety, and content. Participation should be age-appropriate and meaningful,

ensuring children's views help shape digital policies that directly affect them.

4. The Right to Develop Capabilities
   Children's developmental stages and evolving capacities influence the risks and opportunities encountered in the digital world. Policymakers and service providers should tailor digital tools and content to align with children's diverse learning paces and social contexts.
5. The Right to Access Justice and Remedies
   A child-friendly justice system in the digital domain demands legislation with clear sanctions for violations of children's rights online. Children and caregivers must be aware of their legal protections and able to identify violators, collect evidence, and seek redress—even in cross-border digital contexts.
6. Civil Rights and Freedoms in the Digital Context
   - Access to Information: Children have the right to information that is age-appropriate, reliable, and non-discriminatory.
   - Freedom of Expression: Digital platforms should empower children to express their ideas without undue commercial or political manipulation.
   - Freedom of Thought, Conscience, and Religion: Data protection regulations must thwart invasive practices that mine or manipulate children's emotions or beliefs.
   - Freedom of Association and Assembly: Children must be able to form online communities and gather virtually without surveillance by public or private entities.
   - Right to Privacy: Privacy is essential to children's dignity, safety, and broader personal development. Proactive measures must prevent invasive data collection or misuse.
   - Right to Birth Registration and Identity: Digital systems should facilitate—not complicate—legal identity registration processes, especially for marginalized children.
7. Health Rights in the Digital Age
   Innovations in telemedicine and health apps can improve diagnostics and treatment, bridging gaps in maternal, newborn, child, and adolescent health services. At the same time, safeguards must uphold children's data privacy.

8. Educational and Cultural Rights
   Children's access to online learning platforms, e-libraries, and digital cultural resources should be equitable and tailored to their educational stage. Efforts must also address the quality and suitability of online content.
9. Protection from Exploitation and Harmful Practices
   In the digital environment, children are vulnerable to myriad risks—cyberbullying, sexual exploitation, grooming, and manipulative advertising. Strengthened regulations and accountability mechanisms must protect children's physical and psychological well-being.
10. Digital Rights of Vulnerable Children
   Children facing significant adversity—those displaced by conflict, refugees, children in street situations—rely on digital technologies for basic communication, health, and education. Ensuring connectivity and access to digital public services can provide a lifeline for these groups, enabling them to maintain vital family and community ties.

Legislative and Administrative Measures. Paragraph 22 of General Comment No. 25 underscores that "a wide range of legislative, administrative and other measures, including safeguards, are required to ensure that children's rights can be realized and protected in the digital environment." Equally, paragraph 24 reminds States to **review and update** national laws to align with international standards, ensuring that the digital ecosystem upholds children's rights.

Paragraph 25 highlights **cyberbullying and sexual exploitation** as prominent digital threats, also pointing out that children's well-being can suffer if digital content is not adapted to their age, cultural context, or language needs.

Key Stakeholders and Their Roles in Safeguarding Children's Digital Rights. A range of actors can either enable or undermine children's digital rights, depending on the legal, policy, and technical frameworks they adopt.

1. Parents and Caregivers
   - Parents and caregivers often serve as the first line of defense in promoting safe and beneficial use of digital technologies for children.
   - They require access to digital literacy resources, along with the knowledge, skills, and confidence to guide children's online activities effectively.

2. Public Sector (States)
    - Governments bear the primary responsibility for creating and enforcing laws, regulations, and institutional frameworks that protect children's digital rights.
    - They should establish judicial referral mechanisms to handle violations and provide effective support to child victims of digital rights abuses.
    - Regular monitoring and evaluation of relevant programs and services help ensure accountability and continuous improvement.
3. National and International Non-Governmental Organizations (NGOs) and UN Agencies
    - NGOs, UN bodies, and human rights organizations play a critical role in monitoring compliance, raising awareness, and advocating for best practices.
    - Given the cross-border nature of the digital realm, international and regional cooperation is essential, involving states, corporations, and civil society alike.
4. Private Sector
    - Companies offering digital products and services significantly impact children's privacy, autonomy, and well-being.
    - They must ensure transparent policies, provide timely and user-friendly information, and integrate child-safeguarding measures into product design.
5. Commercial Partners
    - Commercial entities engaged in data processing and advertising bear an added responsibility to protect children from exploitative or age-inappropriate content.
    - Preventing the commercialization of a child's attention (e.g., manipulative ads, excessive data tracking) is a key goal.

Addressing Hate Speech and Online Safety. A new **Policy Document, "Countering and Combating Hate Speech Online: A Guide for Policymakers and Practitioners" (July 2023),** outlines key provisions for handling hate speech in the digital age:

1. Respect for Human Rights and the Rule of Law

- Apply human rights standards consistently in content moderation, curation, and regulation of online platforms.

2. Increased Transparency
   - Platforms should be open about their moderation policies and how they handle reports or complaints related to online hate speech.
3. Promotion of Positive Narratives
   - Fostering user engagement and equipping individuals with tools to counter hate speech online can reinforce social cohesion and tolerance.
4. Accountability Mechanisms
   - Strengthening judicial avenues for redress and independent oversight fosters a culture of responsibility and trust.
5. Multilateral Cooperation
   - Governments, international organizations, and civil society need to collaborate closely, given the cross-border nature of hate speech.
6. Context-Sensitive, Knowledge-Based Policies
   - Tailor strategies to local contexts, with emphasis on empowering vulnerable groups and populations to combat online hate effectively.

The Vienna Declaration and Programme of Action (1993). The Vienna Declaration underscores that **the right to development** must be realized to meet the current and future needs of humankind equitably. According to paragraph 11, "everyone has the right to enjoy the benefits of scientific progress and its applications." Crucially, the Declaration calls for **international cooperation** to ensure that rapid scientific and technological developments—including digital technologies—respect human rights and dignity.

International Digital Agenda: Domestic and Global Implications. The digital agenda shapes legal systems at both national and international levels, illustrated by major international policy documents:

1. Okinawa Charter on the Global Information Society (2000)
   - Recognizes information and communication technologies (ICTs) as paramount for 21st-century societal development.

- Paragraph 1 underscores the potential for ICTs to solve complex social and economic problems, spurring growth and enabling social progress.
- Paragraph 9 asserts that "everyone should have access to information and communication networks," and emphasizes eliminating knowledge gaps through international financial institutions such as the Multilateral Development Banks (MDBs), the International Telecommunication Network, UNCTAD, and UNDP.

2. Declaration of Principles: "Building the Information Society—A Global Challenge in the New Millennium" (2003)
   - Links the Millennium Development Goals to the need for equal and broad access to ICT infrastructure, promoting health, education, and environmental sustainability.
   - Affirms the universality, indivisibility, and interdependence of all human rights and fundamental freedoms, citing ICT as a driver of development and economic growth.
   - Paragraph 17 calls for innovative cooperation and partnerships between government bodies and key stakeholders; paragraph 21 stresses "providing universal, ubiquitous, equitable, and affordable access to ICT infrastructure."
   - Recognizes the necessity for continuous learning, ensuring that every individual acquires the skills and knowledge needed to participate effectively in the information society.
3. Principle of Technological Neutrality and Rule of Law
   - Paragraph 39 of the Declaration emphasizes the rule of law in guiding ICT development, while paragraph 44 advocates harmonized international standards to promote an open, human-centered information society.
   - Paragraph 48 highlights the need for multilateral, transparent, and democratic governance of the internet, involving governments, the private sector, civil society, and international organizations.

These documents laid the groundwork for addressing the human rights implications of the Third Industrial Revolution, focusing on issues like internet access, knowledge-sharing, and ICT-driven economic growth.

The digital transformation—spanning from the early internet era to current advancements in AI, big data, and decentralized platforms—has altered the stakes for human rights, particularly for children, who require specialized safeguards and opportunities. International policy documents such as the ***Okinawa Charter*** *and the* ***Declaration of Principles*** *illustrate how global agreements can shape domestic legislation, encourage cross-sector collaboration, and promote equal access to technology.*

However, effectively protecting children's digital rights hinges on more than general statements of principle. It requires a ***multilayered approach****—integrating parental guidance, robust national and international legal frameworks, responsible corporate behavior, and continuous technological innovation. Only through this concerted, multi-stakeholder effort can we ensure that digital technologies serve as powerful catalysts for children's development rather than instruments of exploitation or discrimination.*

From Principles to Actions: The Tunis Commitment Action Plan (2005). The **Tunis Commitment Action Plan (2005)** marked a significant shift from broad principles to concrete actions in the effort to overcome the "digital divide" and manage Internet usage. Notable elements include:

1. Bridging the Digital Divide
   - Paragraph 30 highlights the Internet as a cornerstone of the information society, emphasizing how it has evolved from a research and educational platform to a publicly accessible global tool.
   - Ensuring cost-effective, high-speed connections remains paramount to enabling widespread digital inclusion.
2. Multilingual Internet
   - Paragraph 53 underscores the importance of a multilingual online environment, enabling broader participation across diverse linguistic communities.
3. Data Security and Privacy Concerns
   - Paragraph 40 recognizes the negative implications of rapid technological advances, such as insufficient privacy protections and personal data insecurity. It underscores the need for criminalizing cybercrimes with cross-border implications.

- Existing frameworks (e.g., UNGA resolutions 55/63 and 56/121, the Council of Europe Convention on Cybercrime) guide legislative efforts to mitigate such abuses.

4. Universal Access and E-Government
   - Paragraph 90 stresses the need to ensure both universal and affordable Internet access to achieve development objectives.
   - Subparagraph (j) lays the foundations for e-government applications, promoting more efficient public-sector services.

Interim Conclusions on the Information Society (Circa 2005)

By 2005, global efforts converged on building an inclusive information society founded on the rule of law, reliable legal frameworks to combat cybercrime, and the use of ICT to further development goals (including poverty reduction). Central to this vision was Internet access for all citizens, recognizing ICT's power to expand economic and social opportunities.

Other Pivotal International Instruments. In addition to the Tunis Commitment, numerous other international legal instruments and policy documents address digitalization and its intersection with human and children's rights:

1. UNESCO Charter for the Preservation of the Digital Heritage (2003)
   - Recalls the role of "Information for All" programs and highlights digital heritage as resources related to culture, education, science, governance, and more.
   - Articles 1 and 2 recognize free access (especially for public-domain materials) and promote international cooperation to overcome the digital divide.
   - Article 9 stresses the global, time-transcending nature of digital records, while urging inclusive preservation efforts.
2. "Human Rights and Transnational Corporations and Other Business Enterprises" (2011)
   - Adopts a human-rights-centered view of globalization, emphasizing that ineffective domestic laws can exacerbate inequalities.
   - Calls for an agenda to close governance gaps at the national, regional, and international levels to ensure

transnational corporations do not violate human rights in their operations, including digital and data-related activities.

3. "The Future We Want" (Rio+20 Outcome Document, 2012)
    - Sets the stage for a new era of sustainable development, noting that civil society participation hinges on access to information and digital technologies.
    - Paragraphs 269–275 focus on technology transfer, encouraging partnerships and investment in environmentally sound, inclusive, and development-oriented technologies.
    - Emphasizes the use of space and ground-based monitoring for sustainable development policymaking, acknowledging that rapid technological progress can yield unforeseen social or environmental consequences if not regulated responsibly.
4. UN General Assembly Resolution on Information and Communication Technologies for Development (2013)
    - Confirms the UN's recognition of ICT as a key to addressing global development challenges, with potential benefits for economic growth, social cohesion, and poverty reduction.
    - Simultaneously highlights emerging ICT-related problems, particularly regarding infrastructure gaps and low education levels that hinder meaningful access and usage.

These texts form part of the evolving international legal and policy architecture that shapes digital governance, laying essential groundwork for future action.

Linking ICT to Sustainable Development and Human Rights.

The documents above highlight consistent themes:

- Inclusivity: Overcoming barriers to access so that underserved communities—especially children, women, and marginalized groups—benefit from the digital revolution.
- Rule of Law: Implementing robust national legislation to combat cybercrimes, safeguard personal data, and ensure corporate accountability in the digital domain.
- International Cooperation: Fostering cross-border partnerships and knowledge exchange, which is especially crucial given the global, borderless nature of the Internet.

- Technological Neutrality: Encouraging frameworks that evolve with emerging technologies while preserving core principles of human rights and transparency.
- Sustainable Development: Integrating ICT solutions into broader economic, environmental, and social policies to achieve equitable progress, consistent with the evolving Millennium Development Goals (now the SDGs).

*The **Tunis Commitment Action Plan** and related global instruments reflect the international community's deepening commitment to harness ICT for public good. From preserving cultural heritage to promoting e-government, these initiatives underscore the Internet's transformative potential while acknowledging the pressing need to safeguard **data privacy, human rights,** and **cybersecurity**. Taken together, they illustrate a steadily advancing framework—one that continues to guide national policies, inspire cross-sector collaborations, and pave the way for more nuanced approaches to children's digital rights and broader societal well-being in the digital era.*

UN Council Resolution on the Promotion, Protection, and Enjoyment of Human Rights on the Internet (2014).

This resolution reiterates the foundational principles of the Universal Declaration of Human Rights and the International Covenants on Civil and Political Rights (ICCPR). It reaffirms the equality of offline and online rights, underscoring freedom of expression as a cornerstone of digital rights. Recognizing the Internet's critical role in achieving the Sustainable Development Goals (SDGs), paragraph 2 calls for a human rights-based approach to expanding Internet access—particularly adaptive access for persons with disabilities. Paragraph 8 highlights the importance of addressing new digital security threats that compromise privacy and other human rights.

UN General Assembly Resolution on the Right to Privacy in the Digital Age (2014).

This resolution reinforces the principle that "no one shall be subjected to arbitrary or unlawful interference with his or her privacy, family, home, or correspondence." It calls on states to:

1. Establish or strengthen institutions with **adequate resources** and oversight powers to ensure **transparency and accountability** in state surveillance programs.
2. Guarantee individuals whose privacy is violated by unlawful or arbitrary surveillance **access to effective legal remedies**, consistent with international human rights obligations.

UN Human Rights Council Resolution on the Right to Privacy in the Digital Age (2015).

Continuing the focus on privacy, this resolution states that rights enjoyed offline—particularly the right to privacy—must also be protected online. It led to the appointment of a Special Rapporteur on Privacy, underscoring the growing international consensus on the need for robust privacy safeguards in the digital era.

Sustainable Development and Digital Human Rights.

Sustainable development principles, articulated in documents such as Agenda 21 (1992), the Rio Declaration (1992), and the Millennium Declaration (2000), have also influenced the UN's approach to digital rights. On September 25, 2015, the General Assembly adopted the pivotal resolution A/RES/70/1, "Transforming Our World: The 2030 Agenda for Sustainable Development," which enumerates 17 SDGs. Of particular relevance is Goal 16, focused on peace, justice, and strong institutions, which recognizes access to information as essential—an idea closely tied to digital rights.

High-Level Meeting on the Overall Review of the World Summit on the Information Society (2015).

The outcome document of this meeting reaffirms that ICT can accelerate progress on all 17 SDGs, while acknowledging the darker side of digital technologies. It notes:

- Digital Divide: In 2015, only 43% of the global population—and 41% of women—had Internet access. Around 80% of online content was available in just 10 languages.
- Knowledge-Based Society: A shift is underway whereby information is generated, shared, and harnessed for human development, calling for more equitable Internet access.
- Potential Negative Impacts: Rapid transformations in consumption, social interaction, and time spent online can have adverse environmental consequences and unforeseen outcomes for communities worldwide.

These findings align with global efforts to expand ICT through the Connect the World by 2020 initiative, reinforcing the notion that universal connectivity and digital inclusion are integral to global development.

UN Human Rights Council Resolution on the Right to Privacy in the Digital Age (2018).

Building on the 2014 and 2015 resolutions, the 2018 resolution expands on measures to protect digital privacy. It calls on states to:

1. Respect and Protect Privacy: Enact domestic laws aligning with international standards, especially regarding surveillance, data interception, and collection.
2. Review Legislation: Ensure mass or arbitrary surveillance is subject to impartial oversight, judicial or administrative.
3. Establish Oversight Mechanisms: Provide effective legal remedies for victims of privacy violations.
4. Cooperate with All Stakeholders: Particularly civil society groups.
5. Prevent Violations and Abuses: Implement safeguards to address the evolving landscape of digital privacy threats.
6. Offer Guidance to Businesses: Incorporate the UN Guiding Principles on Business and Human Rights into corporate practices.
7. Encourage Digital Literacy: Expand lifelong learning opportunities for all, so users can protect and exercise their rights online.

The resolution further challenges businesses to:

1. Respect the right to privacy under the UN Guiding Principles on Business and Human Rights.
2. Provide clear information on data collection, storage, and use.
3. Ensure legal, fair, and confidential data processing through administrative, technical, and physical safeguards.
4. Protect human rights in the development and regulation of AI and automated decision-making systems.

UN General Assembly Resolution on Countering the Use of ICT for Criminal Purposes (2019).

This resolution marks the first official UN text identifying artificial intelligence (AI) as a security issue. While ICT and AI offer immense opportunities for state development, they also grant new avenues for criminal activity. The resolution:

- Highlights human trafficking facilitated by ICT as an urgent concern.
- Stresses the importance of international cooperation to combat cybercrime.
- Encourages states to enhance coordination in preventing and investigating digital offenses, including AI-driven abuses.

From the 2014 resolutions on digital privacy and free expression to the 2019 call for heightened coordination against cybercrime, the United Nations has steadily expanded its framework for upholding human rights *in the digital age. Key themes—including **universal Internet access, accountability for surveillance**, and **privacy safeguards**—have become integral to the broader UN agenda on sustainable development and economic progress.*

*In this evolving international context, digital rights are no longer peripheral, but central to discussions about **individual freedoms, global collaboration**, and the ethical use of emerging technologies. As AI continues to reshape social, economic, and governance structures, these foundational UN resolutions provide a roadmap for harmonizing technological innovation with respect for human dignity—ensuring that the digital future is both inclusive and aligned with longstanding human rights principles.*

The Roadmap for Digital Cooperation: A Recent Milestone in Global Digital Governance

The Roadmap for Digital Cooperation, presented in the Report of the UN Secretary-General (29 May 2020), stands as one of the most up-to-date and comprehensive documents on global digital issues. Its analysis reflects an evolving understanding of how digital technologies, especially during the COVID-19 pandemic, simultaneously offer solutions and pose new risks. The Roadmap identifies four main areas for bolstering global digital cooperation:

1. Building an Inclusive Digital Economy and Society
2. Human and Institutional Capacity
3. Human Rights and Their Ability to Actively Shape the World
4. Trust, Security, and Stability

Contextualizing Digital Risks and Opportunities The text underscores that digital technologies do not exist in isolation. While they can drive constructive transformation, they may also exacerbate existing inequalities. The COVID-19 pandemic has accelerated digitization and exposed vulnerabilities that demand urgent attention, including the persistent "digital divide." For instance, 2019 data showed that **87%** of residents in developed countries had internet access, compared to **only 19%**[84] in the least developed countries. Moreover, violations of data privacy worldwide could cost over **US$5**

[84]International Telecommunications Unions (ITU), Measuring Digital Development. Facts and figures 2019 (Geneva, 2019).

trillion[85] by 2024, illustrating the steep potential economic and societal toll[86].

Environmental Concerns also form part of this discussion: up to **20%** of global electricity demand may be attributed to technology-related activities, including data centers that alone account for a third of this consumption.

Human Rights in a Rapidly Evolving Digital Sphere.

Paragraph 28[87] highlights the paradoxical role of digital technologies in promoting, protecting, or suppressing human rights. The Secretary-General's Call to Action for Human Rights stresses that emerging digital tools can enhance surveillance, censorship, and harassment—particularly affecting vulnerable groups and human rights defenders. Paragraph 41 adds that full internet shutdowns or blocking/filtering of services are considered by UN human rights mechanisms to be incompatible with international human rights obligations.

In this context, the Roadmap pinpoints four critical challenges:

1. Data Protection and Privacy
2. Digital Identity
3. Surveillance Technologies, Including Facial Recognition
4. Online Harassment and Violence, Necessitating Content Management and Regulation

Artificial Intelligence: Economic Potential and Societal Risk

AI's global market potential is immense; prior to the COVID-19 outbreak, it was projected to add $4 trillion in value to global markets by 2022[88]. The pandemic has further spurred demands for automated AI-based solutions across multiple sectors. However, AI deployment also introduces serious ethical and security risks—notably the development of lethal autonomous weapons that can make life-and-death decisions without human oversight.

---

[85]Juniper Research, "Business losses to cybercrime data breaches to exceed $5 trillion by 2024," 27 August 2019.

[86]Nicola Jones, "How to stop data centers from gobbling up the world's electricity," Nature, vol. 561, No. 7722 (September 2018).

[87]The Highest Aspiration A Call to Action for Human Rights / URL: www.un.org/sg/sites/www.un.org.sg/files/atoms/files/The_Highest_Asperation_A_Call_To_Action_For_Human_Right_English.pdf (accessed: 25.05.2023).

[88]Gartner, "Gartner says global artificial intelligence business value to reach $1.2 trillion in 2018", 25 April 2018.

In Sum. Overall, universal legal mechanisms lag behind rapidly advancing technologies, offering predominantly declarative principles concerning knowledge economies, privacy, and AI risk mitigation. The UN Roadmap for Digital Cooperation thus stands out as a constructive step toward identifying critical technology-related issues and mobilizing international consensus.

The European Context: Regional Legal Instruments on Digital Rights. Regional documents—in particular those of the Council of Europe—also shape the legal landscape. The Convention for the Protection of Human Rights and Fundamental Freedoms (1950) sets forth, in Article 5, the right to liberty and security of person, anchoring principles relevant to digitalization for the benefit of all.

Council of Europe Frameworks

1. Convention on Cybercrime ETS No. 185 (2001)
   - Establishes substantive and procedural criminal law norms in the digital domain.
   - Defines crimes against confidentiality, misuse of computer facilities, content-based offenses, and infringements related to copyright.
   - Introduces a baseline definition of a "computer system," referring to interconnected devices that automatically process data according to a given program.
2. Parliamentary Assembly Resolution on the Right of Access to the Internet (2014)
   - Recognizes internet access as increasingly vital for democratic participation, freedom of expression, and the enjoyment of human rights.
3. Regulation (EU) 2016/679 (GDPR)
   - Known as the General Data Protection Regulation, it revolutionized data governance by imposing stringent requirements on data processing.
   - Recital 38 observes that children warrant heightened protection around their personal data, as they may not fully grasp the associated risks or safeguards.
   - Recital 58 mandates that any data processing communication targeting children must be written in clear, comprehensible language.
   - Recital 65 affirms the "right to be forgotten," especially pertinent to individuals who consented to data

processing while underage and later seek to erase those digital footprints.
    - Article 8 specifies that a child's personal data processing is lawful only if the child is at least 16 years old; for children under 16, parental or guardian consent is required (though states may set a lower age threshold, down to 13).
4. Declaration on Cooperation in the Field of Artificial Intelligence (2018)
    - Demonstrates Europe's proactive stance on AI governance, including considering Web 4.0 as the next technological phase integrating virtual and physical environments.

Finally, current EU-level work on an "AI Act" proposes frameworks for regulating "high-risk" AI applications (excluding the military domain). Similar to the GDPR's global influence, this regulation could have extraterritorial implications, shaping tech business processes worldwide.

Conclusion. *From the **UN Roadmap for Digital Cooperation** to the **GDPR** and emerging AI regulations, international and regional frameworks are steadily evolving to address the intricate balance between leveraging technology for societal benefit and safeguarding fundamental human rights. While universal legal regulation still lags behind the swift pace of innovation—particularly in AI and data privacy—the initiatives and conventions outlined above reflect growing international momentum toward comprehensive, principled digital governance. As technological frontiers advance, robust cooperation among governments, private-sector stakeholders, and civil society remains vital to ensuring that the digital age enhances, rather than endangers, the rights and well-being of all—especially children and vulnerable populations.*

OSCE Instruments and Their Digital Focus.

In addition to the Council of Europe documents, the Organization for Security and Cooperation in Europe (OSCE) has adopted several notable measures addressing digital transformation and its impact on security and cooperation among member states.

1. Astana Commemorative Declaration on the Path to a Security Community (2010)
    - Paragraph 3 acknowledges that 21st-century security is deeply interconnected; this interdependence

necessitates collective action, including in the sphere of digitalization.

2. Declaration on the Digital Economy as an Engine of Co-operation, Security and Growth (2018)
    - Paragraph 2 observes that the digital economy has become a driver of innovation, competition, and connectivity across the OSCE region.
    - Paragraph 3 points out the mixed effects of digitalization: while technology fosters progress and prosperity, it also introduces new and acute threats.
    - Paragraph 6 underscores the role of digital technologies in fulfilling the 2030 Agenda for Sustainable Development.
    - Paragraph 8 identifies key areas of digital cooperation, including digital innovation in business, bridging the digital divide, promoting international labor standards, and safeguarding human rights in a technology-driven environment.
3. Decision No. 5/18: Development of Human Capital in the Digital Age (2018)
    - Highlights how investing in human capital can bolster socio-economic resilience and reduce corruption, primarily by enhancing digital literacy and fostering a skilled workforce.

Council of Europe Committee of Ministers: Guidelines for Respecting, Protecting, and Fulfilling Children's Rights in the Digital Environment.

The Committee of Ministers issues guidelines advising member states on how to guarantee children's access to the digital environment while safeguarding their rights online. These measures include:

1. Ensuring Access
    - All children, including those with disabilities or living in rural areas, should benefit from equitable infrastructure and high-quality digital services.
2. Freedom of Expression and Access to Information
    - Governments should facilitate the creation of age-appropriate resources while informing children about how to seek help or report abuse.
3. Privacy and Data Protection

- States must inform children of their right to privacy, taking into account age-appropriate consent requirements.
- Default privacy settings should be robust, and profiling of children must be prohibited by law.

4. Education
   - Programs aimed at enhancing children's digital literacy should be embedded in formal educational systems, ensuring broad, equitable access to relevant resources.
5. Safety and Security
   - Policies should anticipate and mitigate children's exposure to online risks, involving all stakeholders (parents, educators, technology providers) in safeguarding efforts.

CIS-Level Initiatives.

In the Commonwealth of Independent States (CIS), the Agreement on Information Interaction of the CIS Member States in the Field of Digital Development of Society (2020) emphasizes:

- Collaborative development of information infrastructure.
- E-government systems and improved public services.
- Efforts to narrow "digital inequality" through shared standards and best practices.

National-Level Examples of Children's Online Protection.

- United States:
  - Children's Online Privacy Protection Act (COPPA) imposes obligations on website operators targeting children under 13.
  - Children's Internet Protection Act (CIPA) requires schools and libraries receiving federal funds to filter inappropriate content.
- United Kingdom:
  - Age Appropriate Design: Code of Practice for Online Services sets 15 technology-neutral standards for online platforms, ensuring default settings minimize data collection while respecting a child's right to information.

Regulating Emerging Technologies and AI. Several jurisdictions have begun addressing generative artificial intelligence. For example, in 2023, one country introduced "Temporary Measures for the Management of Generative Artificial Intelligence Services," obliging AI service providers to label AI-generated content and prioritize societal interests. Additionally, legislation or charter-based approaches in Chile, France, and Spain reflect growing recognition of neuro-rights and mental integrity as areas demanding robust legal protection. The UK, meanwhile, is considering classifying neural data as a special category under GDPR-like frameworks.

Such advancements highlight the increasing importance of multilateral collaboration, aligning national technological standards, and sharing best practices in e-governance for collective digital progress.

Conclusion. The Ongoing Void in Global Digital Human Rights. Despite these myriad initiatives—ranging from OSCE declarations and Council of Europe guidelines to national-level regulations—there is still no universally recognized high-level agreement that cements digital human rights as an indispensable global norm. While many documents acknowledge the importance of privacy, freedom of expression, and data protection in online contexts, they rarely articulate a comprehensive vision of what "digital human rights" truly entail.

In a world increasingly shaped by hyper-connected technologies and instantaneous communication, this absence of a formalized, universal framework evokes a pivotal question: **How do we safeguard human dignity and agency in a reality where algorithms, AI-driven decisions, and rapid data exchange can transform entire societies overnight?** Much like society wrestled with the upheavals of the Industrial Revolutions, we now confront a new epoch where social structures, political systems, and personal identities can be reshaped at digital speed.

Without deliberate, globally coordinated principles that value inclusivity, transparency, and accountability, digital inequalities and power imbalances may deepen, and emerging issues—from AI ethics to neurodata protection—could challenge even the core tenets of human autonomy. The seeds of a solution exist in regional initiatives and thought leadership across multiple organizations. But until we see a world-spanning dialogue on par with past milestones in human rights history—ensuring that digital rights become more than an afterthought—the true promise of technology to elevate humankind risks being overshadowed by fragmentation, exploitation, and mistrust.

In essence, we stand at a crossroads. **Either** we summon the collective will to establish a robust, universal consensus on digital human rights—harnessing technology's capacity for human flourishing—or **we allow** technological advancements to outpace our ethical and legal frameworks, creating a brave new world that lacks a firm foundation in dignity, equality, and justice for all.

Next, we will examine the Global Digital Compact.

2.2. The Global Digital Compact: a historical milestone—and missed openings for digital rights

Context and significance.

Adopted on **22 September 2024** as an annex to the **Pact for the Future**, the **Global Digital Compact (GDC)** is the clearest common statement so far on how the international community intends to steer an open, secure and human-centered digital future. It consolidates years of multistakeholder work stemming from the Secretary-General's *Roadmap for Digital Cooperation* and related processes, and now sits within an intergovernmental agreed framework for action.

Our vantage point. This book treats the GDC as a milestone—but also as a starting line. Compact rightly elevates connectivity, trust, inclusion, responsible data governance and AI guardrails. Yet many commitments remain high-level. To matter for people, principles must resolve into enumerated rights, verifiable standards and enforceable duties for both States and corporations.

What the GDC sets out to do

Close digital divides, accelerate development. Connectivity is framed as essential infrastructure for participation in society and progress on the SDGs. The logic is twofold: universal, affordable access and the capabilities—skills, accessibility, affordability—to use it well.

Open, safe and rights-respecting digital space. The GDC aligns with the long-standing norm that the same rights apply online as offline and engages topics such as internet shutdowns, surveillance, and encryption as determinants of trust and safety.

Fair, interoperable data governance. The Compact signals a path toward equitable, rights-compatible data flows—interoperability, portability, transparency—so data serves people and public value rather than extraction alone.

Responsible AI for human benefit. It calls for global guardrails on AI to prevent discrimination and misuse, while encouraging cooperation on standards and safety across the lifecycle.

Shared responsibility. Implementation is explicitly multistakeholder: governments, companies, technical community and civil society each carry duties in delivery and oversight.

Gains to build on

A common floor for action. By placing digital cooperation in an adopted UN outcome, the GDC creates a reference point for national strategies, regional instruments and platform governance reforms.

Integration with the broader reform agenda. As an annex to the Pact for the Future, the GDC links digital policy with commitments on development, peace and governance, making coordination across ministries and sectors more feasible.

Acknowledgement of systemic risks. Bringing shutdowns, surveillance, encryption and AI governance into one compact underscores that rights, security and innovation are inseparable—and must be addressed together.

Gaps to close

No explicit catalogue of digital rights. The text invokes human-rights language but does not name and define a coherent set of digital human rights (e.g., access and affordability; privacy-by-default; security and strong encryption; data protection and portability; freedom from algorithmic discrimination; explainability and contestability; platform transparency; effective remedy). Without an agreed catalogue, enforcement fragments across jurisdictions.

Limited anchoring in binding law. References to the UDHR, ICCPR, ICESCR and related instruments are not consistently translated into concrete obligations, indicators and remedies. As a result, some duty-bearers may treat digital rights as optional rather than integral to existing law.

Thin cybersecurity architecture. The Compact gestures to safety and trust, but leaves open the "how" of cybersecurity: norms for State behaviour, protection of critical infrastructure, breach reporting and response, encryption-preserving practices, and safeguards for civilians in cyber operations. Given the escalation of shutdowns and cyber incidents globally, this is a material gap.

Corporate accountability in practice. Expectations for private actors (due diligence, impact assessment, remedy) will remain soft without clear auditability, transparency duties and oversight beyond self-regulation.

Reading the operative ideas (without the article numbers)

State duties across the tech lifecycle. The GDC reiterates that States must respect, protect and promote rights in how digital technologies are designed, deployed and governed. The practical corollary is fit-for-purpose institutions: independent oversight, impact assessment and accessible redress.

Harmonizing national law with international norms. States are called to retrofit domestic legislation on data, platforms, cybersecurity and AI so that any limitation on rights is lawful, necessary and proportionate—and so that safeguards and remedies exist in practice.

Engaging platforms and developers. The Compact encourages due diligence and remedy by companies. To be credible, this requires enforceable duties (transparency, audit, records of model governance and data handling) and recognition that certain platforms perform public-interest functions and therefore shoulder heightened responsibilities in a digital commons model.

From principles to protections: how we extend—and operationalize—the Compact

1) Name the rights (Declaration & Convention). We propose a concise Declaration of Global Digital Human Rights with a companion Convention that codifies enforceable rights and duties aligned with existing treaties. The instrument should define the core set listed above and bind both States and corporations to respect, protect and fulfil them, with remedy mechanisms and jurisdictional clarity.

2) Embed "Digital Rights by Design". A techno-legal standard that carries rights into every phase of the technology lifecycle: research and data collection, model training, evaluation (bias audits, red-teaming), deployment (age-appropriate and accessible design; bans on manipulative interfaces), monitoring and

sunset. This is the bridge from declarations to day-to-day engineering and procurement.

3) Build oversight and security. Independent monitoring (periodic reporting, peer review), sector regulators with technical capacity, certification and interoperability schemes, and a modern cybersecurity baseline that protects civilians and critical infrastructure while preserving strong encryption.

Together these moves re-frame today's digital competition—from a race for power and profit to a race for shared prosperity and expanded human capabilities, with data and platforms governed as part of a global digital commons.

Conclusion.

The **GDC** marks a turning point: it records broad agreement that rights must guide digitalisation. But consensus alone does not protect people. Without **explicit digital-rights definitions**, **binding duties** for States **and** companies, and **verifiable safeguards** at platform level, the distance between principle and practice remains wide. The next chapters set out a concrete pathway—**Declaration & Convention**, **Digital Rights by Design**, and governance that treats essential platforms as part of the digital commons—so that the Compact's promise becomes protections in the real world.

*We now turn to the **Pact for the Future** and the unresolved question it raises: what would a universal **right to a digital future** require in legal, technical and institutional terms?*

2.3 Pact for the Future — a vision without a named right to a digital future

Context and significance.

Adopted by the General Assembly in September 2024, the Pact for the Future is the political anchor for renewed multilateralism across peace, development and human rights. It places science, technology and innovation (STI) at the centre of global cooperation, acknowledging both the promise of breakthroughs and the perils of misuse. The Pact's tone is measured optimism: technological progress can widen human freedom—if governance keeps pace and rights remain the compass.

What the Pact gets right.

The text recognizes STI as a force multiplier for the SDGs, urges international cooperation to bridge access gaps, and calls for human-rights considerations in the governance of new and emerging technologies. It also acknowledges that unregulated deployment can entrench inequality, amplify discrimination and undermine civic space. In short, the Pact squarely situates technology within the UN's normative project.

The missing centrepiece.

Yet one element is conspicuously absent: a named right to a digital future. The Pact affirms that technology should advance dignity and development, but it does not articulate a rights-based floor for digital life—no explicit catalogue of digital human rights, and no operational language that would translate principles into enforceable protections for people, especially children. As a result, digital inclusion is framed largely as development policy rather than as a set of legal entitlements with corresponding duties and remedies.

STI and digital cooperation: tools without a rights baseline

The Pact highlights the transformative potential of connectivity, AI, and digital public infrastructure. It points to affordability, accessibility and skills as prerequisites for benefit-sharing. This is the right direction, but it remains programme oriented. Without naming access, privacy by default, data protection and portability, freedom from algorithmic discrimination, explainability and contestability, and effective remedy as rights, States and companies can continue to treat digital accessibility and safety as optional targets rather than obligations.

The same gap appears in risk language. The Pact warns that technology can perpetuate structural inequalities, including gender disparities, but it does not spell out digital-specific harms—mass and targeted surveillance, manipulative design, exploitative data practices, discriminatory automated decision-making, or the militarization of cyberspace. Nor does it define a modern cybersecurity baseline (norms for State behavior, protection of critical infrastructure, breach reporting and response) consistent with human rights and the protection of civilians.

Children and young people: present in vision, absent in digital detail

The Pact rightly elevates children and youth as agents of change, especially in developing regions. It notes deficits in basic services and calls for inclusive opportunities. But it does not set out what a child-centred digital environment requires in practice:

- age-appropriate, rights-respecting product and platform design;
- bans on manipulative interfaces and targeted exploitation;
- data minimisation and privacy by default for minors;
- accessible pathways to advice, complaint and effective remedy;
- curricular commitments to digital literacy and well-being.

Without these elements—aligned with the Convention on the Rights of the Child and General Comment No. 25—children remain the most connected and the most exposed.

Action language that stops short of accountability

Across its action items, the Pact encourages ethical research, inclusive participation and private-sector due diligence. The verbs are largely "encourage" and "promote". Absent are binding expectations for large platforms that structure public discourse and access to opportunity at scale. In today's political economy, corporate systems mediate rights in practice; therefore, duties must attach not only to States but also to companies—with auditability, transparency and enforceable consequences.

From aspiration to architecture: naming the right to a digital future

To align the Pact's vision with real-world protections, this book advances three mutually reinforcing moves:

1) A Declaration and Convention on Global Digital Human Rights. A concise, UN-style instrument that names digital rights (including children's digital rights) and binds duty-bearers—States and corporations—to respect, protect and fulfil them. The text should define the core catalogue (access and affordability; privacy and security with strong encryption; data protection and portability; freedom from algorithmic discrimination; explainability and contestability; platform transparency; and effective remedy) and integrate monitoring and complaints mechanisms.

2) "Digital Rights by Design" as a techno-legal standard. Carry rights into the entire technology lifecycle: research and data collection; model training; evaluation (bias audits, red-teaming); deployment (age-appropriate and accessible design; bans on

manipulative patterns); monitoring and sunset. This bridges Pact-level aspiration with day-to-day engineering, procurement and oversight.

3) Oversight, cybersecurity and the digital commons. Establish independent monitoring (periodic reporting, peer review), sector regulators with technical capacity, interoperability and certification schemes, and a modern cybersecurity baseline that protects civilians and critical infrastructure while preserving strong encryption. Recognise that certain platforms perform public-interest functions and should be governed as part of the digital commons, with heightened duties toward users—especially children and other vulnerable groups.

Conclusion.

The Pact for the Future is a necessary political signal that humanity intends to steer technological change toward dignity and sustainable development. But without a named right to a digital future—expressed as an explicit catalogue of digital rights, embedded in binding law, and carried into code—its promises risk remaining aspirational. By adopting a Declaration and Convention on Global Digital Human Rights, mandating Digital Rights by Design, and building credible oversight and security, the international community can convert the Pact's vision into protections that expand human capabilities, reduce inequality and safeguard the next generation.

The next section turns from political commitments to the legal hierarchy and institutional pathways required to embed these rights within national systems—and within the products and platforms that now mediate everyday life.

2.4. Digital rights of a human and a child in the legislation of the Russian Federation

As an example of how national legislation addresses digitalization, let us consider Russia's legal framework to evaluate whether it meets contemporary requirements for protecting and developing children's rights—especially their digital rights[89]. This

[89]Burianov M.S. Digital human rights as a condition for the effective participation of Russia and other member states of the Eurasian Economic Union in

approach may offer insights for European countries and the United States, particularly as digital development increasingly transcends borders.

Hierarchy of Legal Acts and Digitalization

Russian law consists of:

1. The Constitution of the Russian Federation
2. Federal Constitutional Laws
3. Federal Laws (including Codes, laws amending the Constitution, and laws on ratification or denunciation of international treaties)
4. Subordinate Acts: Presidential Decrees, Government Resolutions, departmental acts, and other measures by federal and regional executive bodies.

Each tier can directly or indirectly influence digital rights, including those of children.

Constitutional Foundations

- Chapter 2 of the Constitution lays a foundational basis for regulating digital rights.
- Article 22 ensures personal freedom and inviolability, while Article 23 upholds privacy and the secrecy of communications, protecting personal honor and reputation. However, Federal Law No. 374-FZ (6 July 2016)—mandating telecommunications providers store large volumes of user data—creates tension with these constitutional guarantees.
- Article 24 prohibits collecting, storing, or disseminating personal data without consent, and compels state bodies to grant individuals access to documents that affect their rights.
- Article 29 guarantees freedom of expression and information, which, in conjunction with Article 17 of the UN Convention on the Rights of the Child, underpins a child's right to access content promoting social, spiritual, and moral development.
- Article 67(4) prioritizes children in state policy, and Article 71(m) highlights the federal government's role in safeguarding individuals, society, and the state against digital threats.

Despite these provisions, Russia lacks robust mechanisms to protect against mass surveillance and the misuse of genetic or medical

---

digitalization 4.0 // Technical and technological problems of service. 2021. No. 2 (56). P. 83-90.

data, as well as enforceable procedures to uphold human rights in the digital realm.

Tension Between Constitutional Principles and Surveillance Laws

Recent legislation has sometimes veered away from constitutional ideals. Federal Law No. 374-FZ obliges internet service providers and mobile operators to store extensive user data, including logs of visited websites, calls, and messages—potentially undermining privacy and due process protections. Observers note the absence of systematic safeguards, such as transparent oversight agencies or clear channels for legal recourse, which challenges the practical realization of constitutional rights.

Impact on Children's Rights

The Family Code guarantees a range of child protections—respect for dignity, measures to shield children from abuse, and provisions for removing a child from perilous conditions. Yet, as the Federal Ombudsman for Human Rights has observed, digitalization implicates virtually all human rights—including children's—but Russia's existing legal instruments often fall short in fully articulating or enforcing digital rights. The COVID-19 pandemic, prompting widespread reliance on online education and services, further exposed legislative gaps.

Russia's constitutional framework acknowledges essential principles—privacy, dignity, and free access to information—that underpin digital rights. However, laws mandating extensive data retention and limited oversight mechanisms risk diminishing these guarantees in practice. The evolving legal landscape highlights a broader challenge: how to reconcile national security and law enforcement objectives with the need to protect and nurture children's digital rights. While Russia's example may offer useful lessons, it also underscores the global imperative for more robust, transparent, and rights-based regulation of digital spaces—particularly where children's interests are at stake.

Legal Framework Protecting Children's Rights in the Digital Age

The Law on Education establishes the right of children to respect for their human dignity and imposes administrative penalties on teaching staff who commit physical or mental violence against students. This ensures a safe and respectful educational environment for all children.

Digital Protection from Exploitation: The Criminal Code holds individuals accountable for physical and sexual violence, including offenses involving minors, as well as crimes against families and children. This framework extends protections into the digital realm, safeguarding children from online exploitation and abuse.

The Child's Digital Right to Information: The cornerstone of information protection for children is the Federal Law No. 436-FZ "On the Protection of Children from Information Harmful to Their Health and Development" (December 29, 2010). This law encompasses various media, including television, radio, the Internet, and the press. It categorizes information based on its potential harm to children's health and development:

1. Intended for children under 6 years of age
2. For children aged 6 years and older
3. For children aged 12 years and older
4. For children aged 16 years and older
5. Prohibited for children

This classification ensures that children are exposed only to age-appropriate content, limiting access to harmful information and regulating the distribution of sensitive material.

The Child's Digital Right to Privacy. Protecting a child's image is governed by Article 152.1 of the Civil Code of the Russian Federation, which stipulates that a citizen's image can only be published or used with their consent. To safeguard children's privacy, their images on television and digital media are typically blurred or obscured unless consent is obtained. According to the Supreme Court of the Russian Federation's Plenum decision (June 23, 2015, No. 25), consent to use an image is a legal transaction under Article 153 of the Civil Code. Only individuals with full legal capacity can grant such consent. For minors:

- Under 14 years old: Legal representatives must consent on their behalf.
- Aged 14 to 18: Minors can consent independently, but only with the permission of their legal representatives.

Additionally, Article 152.1 allows exceptions where consent is not required if the use of the image serves state, public, or other public interests. Personal data processing for minors under 14 can only be consented to by parents, adoptive parents, or guardians, as specified in Article 28 of the Civil Code. Children are legally permitted to register on social networks from the age of 14.

Protection from Developmental Barriers in Information: The Federal Law No. 124-FZ "On the Basic Guarantees of the Rights of the Child in the Russian Federation" (July 24, 1998) mandates state authorities to protect children from harmful information, propaganda, and agitation. This includes safeguarding against:

- National, class, and social intolerance
- Advertising of alcoholic beverages and tobacco products
- Pornographic content
- Information promoting non-traditional sexual relations
- Materials that encourage violence, cruelty, drug addiction, substance abuse, and antisocial behavior

Despite criticisms regarding its legal clarity and the distribution of implementation responsibilities, this law remains crucial for creating a safe information environment for children and adolescents.

Advertising Regulations: Federal Law No. 38-FZ "On Advertising" (March 13, 2006) imposes restrictions on advertising content and distribution methods that may harm a child's health or development. These regulations help prevent children from being exposed to inappropriate or harmful advertising practices.

Information Protection: Federal Law No. 149-FZ "On Information, Information Technologies and the Protection of Information" (July 27, 2006) introduced Article 15.1, which establishes a unified register of domain names and website addresses that disseminate prohibited information in Russia. This includes banning various types of content that could harm minors, ensuring that harmful information is effectively restricted online.

In summary, the Russian legal framework comprehensively addresses the protection of children's rights in the digital environment. By regulating information access, ensuring privacy, preventing exploitation, and restricting harmful content, these laws work together to create a safer and more supportive digital landscape for children.

Enhancements by the Federal Service for Supervision of Communications, Information Technology, and Mass Media

The Federal Service for Supervision of Communications, Information Technology, and Mass Media (Roskomnadzor) has introduced several important additions to the regulatory framework:

1. Addition of Subparagraph (g) to Paragraph 2 of Order No. 152 (03.08.2017): This amendment pertains to the procedure for interaction between the operator of the unified automated information system "Unified Register of Domain Names, Page

Pointers of Sites in the Information and Telecommunications Network 'Internet'," and law enforcement agencies. Despite the introduction of these changes, Order No. 152 has not yet been aligned with Federal Law No. 472-FZ.

2. Federal Law No. 472-FZ (18.12.2018): Titled "On Amendments to Article 15.1 of the Federal Law 'On Information, Information Technology and the Protection of Information' and Article 5 of the Federal Law 'On the Protection of Children from Information Harmful to Their Health and Development'," this law prohibits information aimed at inciting or involving minors in illegal acts that threaten their life or health, or the life and health of others. Such information is subject to blocking and must be transferred to law enforcement agencies through a specialized interaction system as per the procedures approved by Roskomnadzor's Order No. 152. However, it is important to note that Order No. 152 has not yet been updated to comply with Federal Law No. 472-FZ.
3. Legislation on Sexual Exploitation of Minors: The law introduces penalties for forcing a child to perform sexual acts through media, the Internet, or by a group of individuals. Additionally, it addresses crimes committed by individuals previously convicted of offenses against the sexual inviolability of minors. The legislation also allows for the participation of psychologists in the interrogation of minor victims or witnesses during court proceedings. This bill was adopted at the beginning of 2022.
4. Increase in Crimes Against Minors: In 2021, crimes related to the production and distribution of pornographic materials involving minors increased by 30% compared to 2020, with 1,051 cases reported in 2021 versus 809 in 2020. Additionally, there was a 9.5% rise in crimes committed using the Internet for producing pornographic materials under Articles 242, 242.1, and 242.2 of the Criminal Code of the Russian Federation, totaling 2.3 thousand crimes in 2021.

Key Elements of Federal Regulation on Digitalization

The national program "Digital Economy of the Russian Federation" outlines several regulations to govern innovations in areas such as the Internet, big data, artificial intelligence, distributed ledger technologies, and digital twins. The main elements include:

1. **Federal Law No. 34-FZ (18.03.2019)**: Regulates the legal status of "smart contracts" and introduces the concept of digital rights from a private law perspective.
2. **Federal Law No. 63-FZ (15.04.2019)**: Establishes VAT benefits for the export of IT services for businesses.
3. **Federal Law No. 259-FZ (02.08.2019)**: Regulates the procedure for attracting investments through crowdfunding platforms.
4. **Federal Laws No. 436-FZ and No. 439-FZ (16.12.2019), No. 90-FZ (01.04.2020), and No. 268-FZ (31.07.2020)**: Facilitate the transition from paper workbooks to electronic accounting of employee information.
5. **Federal Law No. 480-FZ (27.12.2019)**: Simplifies certain notary procedures, including biometric identification of service recipients.
6. **Federal Law No. 476-FZ (27.12.2019)**: Establishes the institution of a trusted third party and allows the use of "cloud" electronic signatures.
7. **Federal Law No. 478-FZ (27.12.2019)**: Introduces the "register" model for providing public services.
8. **Federal Law No. 122-FZ (24.04.2020)**: Initiates an experiment to implement electronic personnel document management.
9. **Federal Law No. 217-FZ (20.07.2020)**: Allows patents for inventions, utility models, or industrial designs to be issued as electronic documents.
10. **Federal Law No. 259-FZ (31.07.2020)**: Regulates the turnover of digital financial assets.
11. **Federal Law No. 290-FZ (31.07.2020)**: Expands the number of business entities required to offer consumers the option to make payments using national payment instruments (Mir cards).
12. **Federal Law No. 262-FZ (31.07.2020)**: Improves the procedure for the state registration of intellectual activity results by Rospatent.
13. **Federal Law No. 258-FZ (31.07.2020)**: Establishes a regulatory sandbox for individuals engaged in developing and implementing digital innovations, allowing practical application without restrictions from existing regulatory acts.
14. **Federal Law No. 523-FZ (30.12.2020)**: Simplifies procedures and reduces the timelines for developing and updating standardization documents.

15. **Federal Law No. 479-FZ (29.12.2020)**: Enhances the regulation of organizations collecting personal data and conducting remote biometric identification, including the use of corporate biometric systems and expanding the use of a single biometric system for accessing state and municipal services.
16. **Federal Law No. 519-FZ (30.12.2020)**: Establishes conditions for collecting, storing, and processing data using new technologies. It outlines the features of processing personal data and allows data subjects to control the distribution of their personal data, thereby increasing their control over how their data is used.

Digital Rights under the Civil Code

Article 141.1 of the Civil Code of the Russian Federation introduces the concept of digital rights in Russia. Digital rights are defined as "obligatory and other rights named in such capacity in the law, the content and conditions of the exercise of which are determined in accordance with the rules of the information system that meets the characteristics established by law." Key aspects include:

- **Exercise and Disposal**. The exercise, disposal, including transfer, pledge, encumbrance of digital rights by other means, or restriction of disposal, are only possible within the information system without involving a third party.
- **Legal Framework**. Introduced by Federal Law No. 34-FZ (18.03.2019) "On Amendments to Parts One, Two and Article 1124 of Part Three of the Civil Code of the Russian Federation," this concept is distinct from "global digital rights of a person and a child." While digital rights in the Civil Code pertain to the civil law sphere, the latter, still in theoretical development, aim to address the public law sphere, which is crucial for overcoming inequality and digital dictatorships.

The Russian legal framework for protecting children's rights in the digital age is comprehensive and continuously evolving. Through a combination of federal laws, amendments, and regulatory orders, the government seeks to safeguard children from exploitation, ensure their right to information, protect their privacy, and prevent exposure to harmful content. Additionally, the national program "Digital Economy of the Russian Federation" underscores the importance of regulating digital innovations to create a safe and supportive online environment for children and all citizens.

Federal Enhancements to Information and Data Protection

Federal Law No. 149-FZ (27.07.2006, as amended on 02.12.2019) "On Information, Information Technologies and the Protection of Information" regulates the following areas:

- The exercise of the right to search for, receive, transmit, produce, and distribute information.
- The use of information technologies.
- Ensuring the protection of information.

Biometric Data Protection

With the advancement of digital technologies, biometric methods—comprising digital programs that enable automatic identification based on biological and behavioral characteristics such as facial images, fingerprints, retina scans, and voice recordings—have become increasingly prevalent. Protecting biometric data requires a specialized legal regime due to its nature as a unique type of personal data.

Researchers suggest reclassifying biometric data into two categories:

1. Digital Biometrics: Biometric data that leave digital traces.
2. Analog Biometrics: Biometric data that do not leave digital traces.

This classification aims to address and regulate the risks associated with digital biometrics more effectively, ensuring stronger protection and greater individual control over biometric data.

Genetic Data Regulation

Genetic data, which include inherited or acquired genetic characteristics that provide unique information about an individual's physiology or health, as well as relevant biological samples, present another critical area for regulation. In the Russian Federation, the regulation of genetic research and genetic information is still developing. However, there are emerging regulatory trends, such as:

- Decree No. 680 (28.11.2018) "On the Development of Genetic Technologies in the Russian Federation."
- Resolution No. 479 (22.04.2019) "On Approval of the Federal Scientific and Technical Program for the Development of Genetic Technologies for 2019-2027."

These documents outline the direction for accelerating the development of genetic technologies and genetic editing methods, particularly for specific sectors like industry, medicine, and agriculture. There is a recognized need to develop systems for monitoring the risks of genetic technologies and methods to prevent biological emergencies.

Additionally, Federal Law No. 149-FZ and Federal Law No. 152-FZ "On Personal Data" apply to the regulation of genetic data.

Federal Law No. 242-FZ (03.12.2008) "On State Genomic Registration in the Russian Federation" addresses aspects related to the investigation and disclosure of crimes, as well as the collection and processing of genetic information. Legal literature has highlighted challenges in categorizing genetic data as part of biometric data due to the hereditary nature of DNA, which includes information about a person's relatives. This complexity underscores the need for new federal laws to regulate personal biometric and genetic data effectively, especially in the context of increasing privacy concerns during the pandemic.

Amendments to Personal Data Laws

Federal Law No. 152-FZ (03.12.2021) "On Personal Data" introduced significant amendments on March 1, 2021. A new concept, "personal data permitted for distribution," was established, referring to the distribution of personal data to an unlimited number of people. The primary goal of these amendments is to limit the uncontrolled use of data posted on websites and other open sources, enhancing individuals' control over their personal information.

Strategic Development of the Information Society

Decree No. 203 (09.05.2017) "On the Strategy for the Development of the Information Society in the Russian Federation for 2017-2030" was one of the first strategic acts defining Russia's information development vectors. Despite intentions to build an information society, subsequent legislation indicated a shift towards creating a “sovereign Internet.” Federal Law No. 90-FZ (01.05.2019) "On Amendments to the Federal Law 'On Communications' and the Federal Law 'On Information, Information Technology and Information Protection'" raised concerns about restricting Russia’s access to the global information network, potentially hindering participation in rapidly evolving global processes.

International Information Security

Decree No. 213 (12.04.2021) "On Approval of the Fundamentals of the State Policy of the Russian Federation in the Field of International Information Security" is a strategic document outlining Russia's official stance on international information security. It sets three primary goals:

1. Promotion of Russian initiatives on the international stage.
2. Creation of international legal mechanisms to prevent and settle interstate conflicts in the global information space.

3. Organization of interdepartmental cooperation on these issues.

Clause 8 of the Decree identifies main threats to international information security, including:

- The use of information and communication technologies (ICT) in the military-political sphere.
- Technological dominance by individual states in the global information space to monopolize the ICT market.

Paragraph 11 emphasizes Russia's aim to foster regional cooperation within organizations such as the Commonwealth of Independent States (CIS), BRICS, the Collective Security Treaty Organization (CSTO), the Shanghai Cooperation Organization (SCO), the Association of Southeast Asian Nations (ASEAN), and the Group of Twenty (G20). Additionally, it seeks to promote the adoption of the Convention on Ensuring International Information Security and Dialogue on Regulating Information and Communication Technologies by United Nations (UN) member states.

However, the frequent references to combating "terrorism" and "extremism" raise questions about their compliance with the principle of legal certainty. Paragraph 14 outlines the main directions for implementing state policy in international information security, focusing on:

- Countering the threat of ICT use for extremist purposes.
- Preventing interference in the internal affairs of sovereign states through ICT.

The Russian Federation's legal framework for information and data protection, particularly concerning biometric and genetic data, is evolving to address the complexities of the digital age. Recent amendments and strategic decrees reflect an ongoing effort to balance technological advancement with robust data protection and international information security. However, the creation of a "sovereign Internet" and the emphasis on combating extremism and terrorism through ICT highlight the challenges of maintaining legal certainty and ensuring compliance with international norms. Continuous legislative development is essential to protect individual privacy, regulate emerging technologies, and foster a secure and open information society.

Federal Enhancements to Information and Data Protection

Federal Law No. 149-FZ (27.07.2006, amended on 02.12.2019), titled "On Information, Information Technologies and the Protection of Information," governs the following areas:

- The right to search for, receive, transmit, produce, and distribute information.
- The use of information technologies.
- Ensuring the protection of information.

Biometric Data Protection

With the advancement of digital technologies, biometric methods—comprising digital programs that enable the automatic identification of individuals based on their biological and behavioral characteristics such as facial images, fingerprints, retina scans, and voice recordings—have become increasingly prevalent. Protecting biometric data requires a specialized legal framework due to its unique nature as a type of personal data.

Researchers propose reclassifying biometric data into two categories:

1. Digital Biometrics: Biometric data that leave digital traces.
2. Analog Biometrics: Biometric data that do not leave digital traces.

This classification aims to address and regulate the risks associated with digital biometrics more effectively, ensuring stronger protection and greater individual control over biometric data.

Genetic Data Regulation

Genetic data, which include inherited or acquired genetic characteristics that provide unique information about an individual's physiology or health, as well as relevant biological samples, present another critical area for regulation. In the Russian Federation, the regulation of genetic research and genetic information is still in its early stages. However, there are emerging regulatory trends, such as:

- Decree No. 680 (28.11.2018) "On the Development of Genetic Technologies in the Russian Federation."
- Resolution No. 479 (22.04.2019) "On Approval of the Federal Scientific and Technical Program for the Development of Genetic Technologies for 2019-2027."

These documents outline the direction for accelerating the development of genetic technologies and genetic editing methods, particularly for specific sectors like industry, medicine, and agriculture. There is a recognized need to develop systems for monitoring the risks of genetic technologies and methods to prevent biological emergencies. Additionally, Federal Law No. 149-FZ and Federal Law No. 152-FZ, "On Personal Data," apply to the regulation of genetic data.

Federal Law No. 242-FZ (03.12.2008), "On State Genomic Registration in the Russian Federation," addresses aspects related to the investigation and disclosure of crimes, as well as the collection and processing of genetic information. Legal literature has highlighted challenges in categorizing genetic data as part of biometric data due to the hereditary nature of DNA, which includes information about a person's relatives. This complexity underscores the need for new federal laws to effectively regulate personal biometric and genetic data, especially in the context of increasing privacy concerns during the pandemic.

Amendments to Personal Data Laws

Federal Law No. 152-FZ (03.12.2021), "On Personal Data," introduced significant amendments on March 1, 2021. A new concept, "personal data permitted for distribution," was established, referring to the distribution of personal data to an unlimited number of people. The primary goal of these amendments is to limit the uncontrolled use of data posted on websites and other open sources, thereby enhancing individuals' control over their personal information.

Strategic Development of the Information Society

Decree No. 203 (09.05.2017), "On the Strategy for the Development of the Information Society in the Russian Federation for 2017-2030," was one of the first strategic acts defining Russia's information development vectors. Although these trajectories aimed to build an information society, subsequent legislation indicated a shift towards creating a "sovereign Internet." Federal Law No. 90-FZ (01.05.2019), "On Amendments to the Federal Law 'On Communications' and the Federal Law 'On Information, Information Technology and Information Protection'," raised concerns about restricting Russia's access to the global information network, potentially hindering participation in rapidly evolving global processes.

International Information Security

Decree No. 213 (12.04.2021), "On Approval of the Fundamentals of the State Policy of the Russian Federation in the Field of International Information Security," is a strategic document outlining Russia's official stance on international information security. It sets three primary goals:

1. Promotion of Russian Initiatives on the International Stage: Enhancing Russia's influence and participation in global information security discussions.

2. Creation of International Legal Mechanisms: Developing frameworks to prevent and resolve interstate conflicts in the global information space.
3. Organization of Interdepartmental Cooperation: Facilitating collaboration among various government departments to address information security challenges.

Clause 8 of the Decree identifies the main threats to international information security, including:

- The use of information and communication technologies (ICT) in the military-political sphere.
- Technological dominance by individual states in the global information space to monopolize the ICT market.

Paragraph 11 emphasizes Russia's aim to foster regional cooperation within organizations such as the Commonwealth of Independent States (CIS), BRICS, the Collective Security Treaty Organization (CSTO), the Shanghai Cooperation Organization (SCO), the Association of Southeast Asian Nations (ASEAN), and the Group of Twenty (G20). Additionally, it seeks to promote the adoption of the Convention on Ensuring International Information Security and Dialogue on Regulating Information and Communication Technologies by United Nations (UN) member states.

However, the frequent references to combating "terrorism" and "extremism" raise questions about their compliance with the principle of legal certainty. Paragraph 14 outlines the main directions for implementing state policy in international information security, focusing on:

- Countering the threat of ICT use for extremist purposes.
- Preventing interference in the internal affairs of sovereign states through ICT.

Role of Comprehensive Child Safety Strategy

Decree No. 358 (17.05.2023), "On the Strategy for Comprehensive Child Safety in the Russian Federation for the Period up to 2030," defines the main directions, goals, and objectives of the policy for enhancing child safety. One of the key directions is the creation of a modern, secure infrastructure to protect children in digital environments.

Digital Rights and International Standards

In conclusion, the Russian legal concept of "digital rights" is enshrined in Article 141.1 of the Civil Code of the Russian Federation, within the framework of private law regulation. This contrasts with international practices, where digital rights are typically considered from the perspective of public law, focusing on children's rights in the context of digital technology development. Fundamental provisions on children's rights, such as the Convention on the Rights of the Child, play a crucial role in regulating digitalization.

International treaties and domestic acts have established some directions for the future digital rights of children, ensuring their right to access technological benefits without harm. Based on an analysis of key international legal treaties and domestic regulations, we highlight several essential digital rights of children:

1. Right to Life, Survival, and Development. Especially critical in crisis situations, technology's role in early childhood development must be considered in its design, purpose, and use to support cognitive, emotional, and social growth.
2. Right to Non-Discrimination. Ensuring all children have equal, effective, and meaningful access to the digital environment, including safe access in public spaces and investment in policies that promote affordable and wise use of digital technologies.
3. Right to a Child's Opinion. Involving children in the development of legislation, policies, programs, services, and training related to their digital rights, listening to their needs, and considering their views.
4. Right to Develop the Child's Capabilities. Recognizing that the risks and opportunities in the digital environment vary with age and developmental stages.
5. Right to Access Justice and Legal Remedies. Addressing implementation challenges such as the lack of specific legislation, difficulty in obtaining evidence or identifying violators, and awareness gaps among children and their guardians.
6. Civil Rights and Freedoms in the Digital Context. Including the right to access information, freedom of expression, freedom of thought, conscience, and religion, freedom of association and peaceful assembly, and the right to privacy.
7. Health Rights in the Digital Context. Facilitating access to health services and information, improving diagnostics and treatment, and ensuring adequate nutrition for mothers, newborns, children, and adolescents.
8. Educational and Cultural Rights. Providing opportunities for modern education and access to cultural values through digital technologies.

9. Right to Protection from Exploitatio. Safeguarding children from all forms of exploitation that harm their well-being in the digital environment.
10. Digital Rights for Vulnerable Children. Ensuring that vulnerable children, including those in armed conflict, internally displaced, migrant, asylum-seeking, refugee, unaccompanied, street children, and those affected by natural disasters, have access to vital information, maintain contact with their families, and receive essential services through digital means.

The Importance of Mechanisms and Norms

Despite these protections, creating positive conditions for the comprehensive development of children amidst rapid technological innovations remains challenging. Therefore, developing mechanisms and norms to realize children's digital rights is of paramount importance.

The Role of Artificial Intelligence in Digital Rights

In July 2023, the UN Security Council discussed artificial intelligence (AI) for the first time, addressing its challenges and opportunities. António Guterres highlighted the unprecedented speed of technological advancement, comparing it to the invention of the printing press but noting that AI technologies like ChatGPT reached 100 million users in just two months.

We concur with the Council's view on the need to introduce principles into international law concerning AI:

- Openness. AI should support freedom and democracy.
- Responsibility. AI must comply with the rule of law and human rights.
- Safety and Predictability. Protecting property rights, privacy, and national security.
- Public Trust. Ensuring AI systems are trustworthy and their critical systems are protected.

Additionally, it is essential to develop digital human rights within the context of AI, focusing on technologically implementing human rights and expanding opportunities.

Regional and Global Cooperation

Implementing AI is already a priority for every Russian region and department. Therefore, securing the digital rights of individuals and children at the federal level is crucial. This involves developing digital rights principles within modern technological systems and fostering regional cooperation within organizations such as the CIS, BRICS, SCO, and the EAEU. Extensive work has been undertaken to assess the need for regional and global cooperation in this area, emphasizing the importance of integrated efforts to protect digital rights effectively.

The Russian Federation's legal framework for information and data protection, particularly concerning biometric and genetic data, is evolving to address the complexities of the digital age. Recent amendments and strategic decrees reflect ongoing efforts to balance technological advancement with robust data protection and international information security. However, the creation of a "sovereign Internet" and the emphasis on combating extremism and terrorism through ICT highlight the challenges of maintaining legal certainty and ensuring compliance with international norms. Continuous legislative development is essential to protect individual privacy, regulate emerging technologies, and foster a secure and open information society for all, especially children.

Laureate Projects and the Vision for Techno-Legal Platforms.

In January 2021, the concept behind these techno-legal platforms became a laureate of the VI International Competition for the Best Scientific Work/Project "Eurasian Integration: Youth Dimension" in St. Petersburg. A month later, in February 2021, this framework for recognizing and enforcing digital rights took first place at the International Competition of Scientific Research and Project-Creative Works of Young Scientists of Eurasia, "Science and Creativity: Dialogue and Development" (also in St. Petersburg), specifically in the category "Legal Aspects of the Activities of the Eurasian Economic Union."

The concept continued to gain momentum in October 2021, winning the VII International Competition for the Best Scientific Work—"Eurasian Integration: Youth Dimension"—as part of the IX International Forum "Eurasian Economic Perspective," hosted by the St. Petersburg State University of Economics in partnership with the Eurasian Economic Commission. The winning project, titled "Techno-Legal Platforms as a Foundation for Constructing an Innovative State and Integration Institutions within the Eurasian Economic Union for Sustainable Development," advanced the idea that synergy among legal frameworks, technological solutions, and public policy could accelerate sustainable digital growth.

Similarly, in December 2021, the concept of techno-legal platforms and the notion of a new **ESG 4.0** secured first place at the "Young Analyst of Eurasia" international competition with the research "Principles of Techno-Legal Platforms and New ESG 4.0 in the Context of Creating Global Governance." Across more than twenty international and domestic competitions from 2017 to 2024, these integrative approaches have consistently garnered recognition—evidencing their relevance for global and regional cooperation in digital human and child rights, particularly when viewed against the backdrop of the UN Sustainable Development Goals and fast-evolving AI technologies.

Toward a New Social Contract for Human and Children's Digital Rights.

In an era of rapid digital transformation—a phenomenon described by Klaus Schwab as the "Fourth Industrial Revolution"—merely safeguarding children's rights is insufficient. Far more crucial is proactive engagement to realize those rights. This means constructing robust legal and technical mechanisms so that young people can flourish in digital societies. Digital rights for children should become the cornerstone of a new social contract, emphasizing equitable development and justice in a technologically driven age.

Seven Key Stakeholder Groups in the Ecosystem.

1. Public Sector and International Organizations: Digital Legislators and Policymakers
Government bodies and international institutions (e.g., UN, UNICEF, WHO) shape the legislative and regulatory landscape of digital human and child rights. Their proactive initiatives—aimed at reducing digital inequality and ensuring consistent implementation—are vital for integrating digital rights into broader national and international development strategies.
2. Private and Technology Sector: Digital Innovators and Developers
Tech companies, startups, service providers, and venture funds serve as the engine of innovation. By developing secure, privacy-respecting platforms, they empower individuals—including children—to leverage digital tools for learning, self-expression, and civic engagement. Their leadership also extends to digital inclusion initiatives and investing in digital literacy.
3. Tech Sector and Cybersecurity Agencies: Digital Defenders and Technical Implementers
Cybersecurity agencies and IT consultants set the standards for robust digital infrastructures. Their responsibility is twofold: protect users from cyber threats and reinforce rights such as privacy and autonomy in secure digital domains. They embody

the ideal of "digital sovereignty" while maintaining global connectivity.

4. Legal and Human Rights Sector: Digital Lawyers and Advocates Law firms, NGOs, and human rights organizations provide the legal frameworks and redress mechanisms necessary for enforcing digital rights. Going beyond litigation, they champion forward-looking legislation and international engagement, ensuring that fundamental values—like equality and dignity—remain central to technological progress.
5. Education and Research Sector: Educators and Innovators Universities and research centers perform a dual role: embedding digital technologies into their curricula while critically examining societal effects. By developing digital literacy programs and producing evidence-based policy advice, they help construct secure, child-friendly digital ecosystems aligned with educational objectives.
6. Media Sector and Platforms: Promoters of Ideas and Moderators of Dialogue Traditional and digital media shape public consciousness around digital rights, promoting safe online behavior and driving global dialogue. Through responsible reporting and user education, they elevate conversations about ethics, oversight, and the protection of vulnerable populations—especially children.
7. Global Community: Advocates of Interests Civil society groups, parent associations, and grassroots movements form the final stakeholder tier. Through online petitions, social campaigns, and community-building efforts, they cultivate digital ethics and nurture an environment where children's rights can thrive. Their collective voice ensures that, beyond institutional mandates, everyday users can safeguard and expand their digital freedoms.

Conclusion of Chapter 2: Bridging the Gap Between Innovation and Dignity.

These seven stakeholder groups collectively hold the key to fostering a just, safe, and forward-thinking digital society. Yet, as Yuval Noah Harari has pointed out in his explorations of humanity's future, societies often adopt transformative technologies faster than they can fully grasp their long-term consequences. Despite the proliferation of local and regional laws and the acknowledged importance of digital rights, the global legal framework remains fragmented—especially when viewed through the lens of the Fourth Industrial Revolution.

This gap underscores the imperative to develop comprehensive legal norms, protocols, and regulatory instruments that can evolve in tandem with cutting-edge technologies. The lack of a universal standard jeopardizes not only the ideals of justice and equality but also the potential for human-centered innovation that respects personal autonomy and dignity.

By synthesizing legal, technological, and ethical principles through **techno-legal platforms** and forging **ESG 4.0** strategies, nations and organizations alike can build pathways to sustainable, rights-based digital development. In an interconnected era where AI transcends geopolitical borders, forging international consensus on digital human and child rights is both a moral obligation and a pragmatic necessity. It is only through such dedicated, multi-stakeholder collaboration that we can ensure the digital era evolves into an inclusive renaissance—one that safeguards our children's futures and elevates the human experience.

Since the adoption of the Universal Declaration of Human Rights (1948), the international human-rights regime has moved from a values-based declaration to binding codification: the International Bill of Human Rights—the 1966 Covenants on Civil and Political Rights (ICCPR) and on Economic, Social and Cultural Rights (ICESCR)—together with the Vienna Declaration and Programme of Action (1993) affirmed the universality and indivisibility of rights. Further elaboration proceeded through specialized treaties on non-discrimination and the protection of vulnerable groups: the International Convention on the Elimination of All Forms of Racial Discrimination (1965, CERD); the Convention on the Elimination of All Forms of Discrimination against Women (1979, CEDAW); the Convention against Torture and Other Cruel, Inhuman or Degrading Treatment or Punishment (1984, CAT); the Convention on the Rights of the Child (1989, CRC); the International Convention on the Protection of the Rights of All Migrant Workers and Members of Their Families (1990, ICMW); the International Convention for the Protection of All Persons from Enforced Disappearance (2006, CPED); and the Convention on the Rights of Persons with Disabilities (2006, CRPD). In parallel, collective and solidarity rights took shape—the Declaration on the Right to Development (1986), ILO Convention No. 169 on Indigenous and Tribal Peoples (1989), and the UN Declaration on the Rights of Indigenous Peoples (2007)—while regional regimes (the European Convention on Human Rights, 1950; the American Convention on Human

Rights, 1969; and the African Charter on Human and Peoples' Rights, 1981) added judicial protection and contextualization.

With the onset of the Fourth Industrial Revolution, "offline" rights were extended to the online environment primarily through resolutions (HRC 20/8, 2012); privacy was reframed for the digital age (GA 68/167 and subsequent texts); and the WSIS framework (Geneva 2003; Tunis 2005) linked access, development and rights. Sectoral standards strengthened legal guarantees: the Council of Europe's Convention 108 and its modernized "108+" (1981/2018); the Budapest Convention on Cybercrime (2001); and the UNESCO Universal Declaration on Bioethics and Human Rights (2005). The business-and-human-rights package translated principles into due-diligence and remedy processes: the UN Guiding Principles on Business and Human Rights (2011), the updated OECD Guidelines for Multinational Enterprises (2023), the ILO Declaration on Fundamental Principles and Rights at Work (1998, amended 2022), and the ILO Tripartite Declaration for MNEs (2017). Recent advances include recognition of the right to a clean, healthy and sustainable environment (HRC 48/13; GA 76/300) and the consolidation of AI ethics and law (UNESCO Recommendation on the Ethics of AI, 2021; Council of Europe Framework Convention on AI, 2024; GA res. 78/265, 2024). The overall trajectory is from principles to enforceable obligations and "rights-by-design" across digital infrastructures. In 2024, the Global Digital Compact was adopted; in our view, this is only a starting point, and the principal strategic instrument still ahead should be a **Declaration of Global Digital Human Rights**—to define the core catalogue of digital rights, set a human-centered vector for digital transformation, enable a proactive, rights-implementing digital state that mitigates harm and abuse, and lay the groundwork for decentralized global governance.

CHAPTER 3. PROBLEMS AND PROSPECTS OF REALIZING THE DIGITAL RIGHTS OF HUMAN AND THE CHILD

3.1. Problems of implementing the digital rights of the child

In this chapter, we delve into the complexities and opportunities surrounding the implementation of digital rights for children. We will explore the challenges that hinder these rights and outline a systematic approach to establishing new regulatory frameworks. Today, technology is not just a tool but an integral part of human experience, shaping how we perceive, hear, act, and communicate. The rapid integration of digital technologies into social relations amplifies the importance of safeguarding and advancing children's digital rights.

Harnessing Technology for Sustainable Development. Echoing the vision of UN Secretary-General António Guterres, digital technologies hold the promise of significantly advancing sustainable development. To ensure these technologies contribute positively, Guterres advocates for a **Global Digital Compact**[90]—a comprehensive framework for an open, free, inclusive, and secure digital future for all, including children. He emphasizes that "a human-centered digital space starts with protecting free speech, freedom of expression, and the right to online autonomy and privacy," placing the onus on governments, tech companies, and social media platforms to uphold these principles.

Addressing the Rise of Artificial Intelligence. In July 2023, Guterres addressed the Security Council, focusing on the transformative potential of artificial intelligence (AI). He highlighted the need for the international community to adapt to new technologies that profoundly impact society and the economy. Guterres called for the United Nations to establish new international rules, sign new treaties, and create global agencies dedicated to AI governance. Among his proposals is the creation of the **New Agenda for Peace**, a think tank designed to guide member

---

[90]UN chief: Need to help 3 billion people on planet get internet access / URL:https://news.un.org/ru/story/2022/11/1434817(date accessed: 06.06.2023).

states in managing AI responsibly. Additionally, he urged the conclusion by 2026 of negotiations on a legally binding instrument to ban lethal autonomous weapons systems—machines capable of making life-and-death decisions without human oversight, in violation of international humanitarian law.

Gaps in Current Approaches to Digital Human Rights. Despite these visionary initiatives, there remains a significant gap in the scientific and theoretical frameworks necessary to establish robust norms for protecting and implementing digital human rights, particularly for children. Current approaches primarily focus on ensuring the reliability of public information, enhancing free speech, and safeguarding privacy in personal life. However, they often overlook the broader impact of **Industry 4.0** technologies—ranging from AI and big data to biotechnology and the Internet of Things (IoT)—on a wider array of universal human rights, including those of children.

Pioneering a Declaration of Global Digital Human Rights.

Recognizing these deficiencies, I initiated in 2020 the development of a **Declaration (and Convention) of Global Digital Human Rights**. This project, which incorporated input from stakeholders across 30 countries and was endorsed by 70 youth organizations[91], was envisioned as a new global social contract for the digital age. Two years after its inception, elements of this declaration began to take shape, signaling the beginning of its implementation and underscoring its relevance in today's digital landscape.

The Acceleration of Digital Integration.

Technology's relentless advancement into every facet of social relations necessitates a comprehensive examination of children's digital rights. The speed at which digital transformations occur often outpaces our ability to legislate and regulate effectively, creating a precarious environment where children's rights can be both empowered and endangered by the very technologies designed to enhance their lives.

Key Challenges in Upholding Children's Digital Rights

1. **Privacy and Data Protection**: As children engage more with digital platforms, the risk of data breaches and unauthorized data collection escalates. Ensuring robust data protection

[91]Burianov M.S. Digital human rights in the context of global processes: theory and practice of implementation: monograph - Moscow: RUSAINES, 2022. - 148 p. - P. 120.

mechanisms is paramount to safeguarding their personal information.

2. **Online Safety and Harassment**: The digital realm can expose children to cyberbullying, exploitation, and other forms of online harassment. Developing effective content moderation and protective measures is essential to create a safe online environment.
3. **Access and Inclusion**: Bridging the digital divide remains a critical challenge. Ensuring equitable access to digital technologies and the internet for all children, regardless of socioeconomic status or geographic location, is fundamental to realizing their digital rights.
4. **Digital Literacy and Education**: Empowering children with the knowledge and skills to navigate the digital world responsibly is crucial. Digital literacy programs must be integrated into educational curricula to prepare children for the complexities of the digital age.
5. **Ethical AI and Technological Governance**: The deployment of AI and other advanced technologies in children's lives necessitates ethical oversight to prevent biases, discrimination, and unintended consequences that could infringe upon their rights.

Towards a Systemic Implementation of Global Digital Rights. Addressing these challenges requires a holistic approach that integrates legal, technical, and educational strategies. Establishing a robust system for implementing new regulations involves:

- **Legislative Reform**: Updating existing laws and enacting new ones to reflect the realities of the digital age, ensuring that children's rights are explicitly protected within digital frameworks.
- **Multi-Stakeholder Collaboration**: Engaging governments, tech companies, educators, and civil society in a coordinated effort to develop and enforce digital rights protections.
- **Innovative Regulatory Mechanisms**: Creating adaptive and forward-looking regulatory bodies capable of responding to the rapid pace of technological change.
- **Global Cooperation**: Fostering international agreements and standards that transcend national boundaries, ensuring consistent protection of digital rights worldwide.

I have been deeply involved in laying the groundwork for why it is crucial to develop international laws on digital human rights. As part of the Commonwealth of Independent States (CIS) group, I have actively participated in global events related to Generation Connect (GC). My ideas on digital rights were prominently featured in the 2022 manifesto, *Generation Connect Youth Call to Action: My Digital Future*[92]. Below, we highlight key points from the manifesto that were incorporated into the final version concerning digital rights.

Modernizing Policy and Governance for the Digital Age. Many current policymaking methods, governance models, and norm enforcement mechanisms require significant updates to keep pace with the digital age. These can be enhanced by leveraging digital tools and the internet to boost social interaction, decision-making, and leadership. This ensures the meaningful inclusion of youth and other underrepresented groups. Decisive action is needed to meet the following requirements by 2025:

- **Digital Policy Enhancements.**
  - **Point c:** Governments must respect human rights online by developing systems that ensure safe navigation, free expression, consent for online transactions, and access to social and economic opportunities through digital technologies.
  - **Point e:** Governments should support the development and enforcement of industry codes of conduct to better protect the data, integrity, and digital rights of young people, children, and all online users.
- **Transforming Culture and Communities.**
  - **Item d:** Develop and implement both online and offline awareness-raising campaigns on key topics related to youth engagement in the digital economy and society. This includes protecting children's privacy and rights online, combating online violence and addiction, promoting the ethical use of digital media and technologies, and ensuring the accuracy and reliability of online information.
  - **Item e:** Collaborate with young people to create and execute awareness campaigns addressing issues such as

[92]2022 Generation Connect Youth Call to Action My Digital Future / URL:https://www.itu.int/generationconnect/wp-content/uploads/2022/06/GenerationConnectYouthCallToAction2022.pdf(date accessed: 14.06.2023).

adolescent rights, mental and reproductive health, domestic violence, and social challenges related to disability.

Recognizing Youth and Children as Equal Stakeholders.

The inclusion of such initiatives worldwide signifies that young people and children are being recognized as key stakeholders alongside states in shaping the future global digital treaty. Earlier, in 2020, I proposed and developed a project for the cyber socialization of children—a specialized device (a children's phone) supported by the Global Law Forum. This project aimed to implement certain digital rights for children and won the final phase of the international youth forum "Shaping the Future Together[93]," supported by the UN Information Center in Moscow.

Contributions to Global Digital Access.

In August 2021, as part of the Global Shapers (World Economic Forum) initiative, I contributed to the **Davos Lab: Youth Recovery Plan Insight Report**[94]. My proposals for the Global Digital Human Rights project were incorporated into the Digital Access section. The report emphasizes that access to digital technologies is a fundamental human right, noting that disparities in digital access exacerbate inequality. Approximately 89% of respondents believe that digital access should be a basic human right, highlighting the need for sanctions against institutions that block internet access and deprive individuals of basic rights and freedoms.

Adapting to Technological Advancements

Since these conclusions were drawn, the tech industry has experienced significant shifts. Trends such as the increasing use of technology for military purposes and the rise of digital surveillance have become more prominent. Simultaneously, innovations like generative artificial intelligence, Web 3.0 concepts, and metaverse startups have gained substantial momentum. These changes underscore the critical importance of envisioning the future of digital human rights, particularly from the perspective of younger generations.

Defining and Implementing Digital Human Rights

---

[93]Technology for Good / URL:http://www.futurible.space/media/attachments/website_almanacwork/157/Tekhnologii_uluchshayuschie_zhizn_6.pdf(date accessed: 14.06.2023).

[94]Davos Lab: Youth Recovery Plan Insight Report August 2021 / URL:https://www3.weforum.org/docs/WEF_Davos_Lab_Youth_Recovery_Plan_2021.pdf(date accessed: 14.06.2023).

To address these evolving challenges, we first define the concept of implementing digital human rights and systematically examine the main stages through which implementation problems emerge. According to A.G. Tarasova, a dissertation author, the mechanism for implementing human rights can be defined as a "legally provided set of actions by a person as a right holder and other subjects with the purpose of obtaining a social benefit mediated by one or another of their rights, as well as the use and disposal of the benefit. [95]"

Challenges in Implementing Children's Digital Rights

Several key challenges impede the effective implementation of children's digital rights:

1. **Legislative Inconsistencies:** Variations in the content of children's rights across different branches of legislation create confusion and weaken protection mechanisms.
2. **Lack of Effective Mechanisms:** There is a scarcity of specific and effective mechanisms for protecting and providing for children's rights.
3. **Underdeveloped Legal Norms:** Insufficient development of norms and provisions within legal acts hampers the effective implementation of children's rights.
4. **Digital Transformation Barriers:** Challenges include a lack of awareness about digital technologies and their potential risks, limited access to technologies, and insufficient digital literacy among stakeholders. According to UNICEF, barriers such as inadequate internet access, high connection costs, and poorly designed digital services exclude many families, especially those with children, from utilizing essential digital services[96].

The **Order of the Government of the Russian Federation dated 20.05.2023 No. 1315-r** outlines general challenges of digital transformation in Russia that may impact access to the digital benefits of Industry 4.0, including technological imperfections in production

---

[95]Tarasova A.G. Legal procedures and implementation of human rights: theoretical and legal aspect: dis. ... candidate of legal sciences. - M. 2012. - 206 p. - P. 13.

[96]Towards a child-centred digital equality framework. UNICEF/URL:https://www.unicef.org/globalinsight/media/2966/file/UNICEF-Global-Insight-Towards-a-child-centred-digital-equity-framework.pdf(date accessed: 01.06.23).

systems across various industries and violations of infrastructure security, including information security issues.

F. M. Rudinsky identifies the following stages of human rights implementation:

1. **Recognition in International Law (IL).**
2. **Implementation at the Domestic Level.**
3. **Preparation for Implementation.**
4. **Implementation.**
5. **Protection.**

Modern researchers also emphasize theoretical and legal development as foundational stages of human rights implementation.

Systemic Implementation of Children's Digital Rights. Implementing children's digital rights involves creating normative, educational, and digital solutions[97]. This includes:

- **Normative Solutions:** Establishing legal frameworks that explicitly protect digital rights.
- **Educational Solutions:** Developing programs focused on digital hygiene and literacy.
- **Digital Solutions:** Implementing technological tools, such as secure codes, machine programs, and cryptographic standards, to ensure children can access social benefits and develop without technological barriers.

Identifying Implementation Problems.

Systematically identifying problems in implementing children's digital rights involves examining:

1. **Scientific and Theoretical Level:** Defining "digital rights of the child" and related concepts like "digital control," "digital militarization," and "digital inequality."
2. **Normative and Legal Level:** Addressing inconsistencies and gaps in legislation.
3. **Law Enforcement and Techno-Legal Platforms:** Developing and enforcing standards and educational programs for digital hygiene and literacy.

---

[97]Burianov M. S. Deconstruction of Law against the Background of Globalization and Planetary Risks // Intellectual Culture of Belarus: Cognitive and Prognostic Potential of Social and Philosophical Knowledge: Proceedings of the Fourth Int. sci. Conf. (November 14–15, 2019, Minsk). In 2 vols. Vol. 2 / Institute of Philosophy of the National Academy of Sciences of Belarus; editorial board A. A. Lazarevich (chairman) [and others]. Minsk: Four Quarters, 2019. – Pp. 38–41.

*

The Regulatory Gap: Missing Legal Protections for Children's Digital Rights. At the heart of the issue lies a significant regulatory gap. There is a lack of comprehensive laws that enshrine the fundamental digital rights of children and establish mechanisms for their implementation. This deficiency poses substantial risks to the development and protection of children's rights in the digital age. Without robust legal frameworks, children are vulnerable to various digital risks, especially those associated with Industry 3.0 technologies such as the Internet, personal computers, and smartphones.

Understanding Digital Risks: The EU Kids Online Framework[98].

To better understand these risks, we can refer to the classification developed by the European Network for Research on Children and the Internet (EU Kids Online). This framework categorizes digital risks into three main areas:

1. **Content Risks**. These involve exposure to mass-produced content through access to mass media, social networks, and other digital platforms. Examples include exposure to violent or inappropriate material, hate speech, and information overload, which can lead to issues like "clip thinking" where individuals struggle to process large amounts of information effectively.
2. **Contact Risks**. These occur when children engage in activities initiated by adults. This includes interactions that may not always be positive, such as harassment, stalking, or radicalization efforts aimed at persuading children to participate in harmful behaviors.
3. **Behavioral Risks**. Contrary to the popular notion of children as passive victims in the digital environment, behavioral risks highlight the active role children may play online. This includes digital violence, humiliation of peers, and the creation or dissemination of hateful content. Authors like Barbovschi and Dreier argue that this perspective challenges the myth of the innocent child, recognizing that children can also be perpetrators of digital harm, often influenced by flawed platforms and algorithms.

[98]Fourie, L. (2020). Protecting children in the digital society. In J. Grobbelaar & C. Jones (Eds.), Childhood vulnerabilities in South Africa: Some Ethical Perspectives (1st ed., pp. 229–272). African Sun Media.

The Evolution to Web 3.0: New Challenges and Risks. As we transition from Industry 3.0 to Web 3.0, the landscape of digital risks evolves. Web 3.0, characterized by augmented reality (AR), virtual reality (VR), the Internet of Things (IoT), and blockchain technologies, introduces new vulnerabilities and cybersecurity challenges. These advanced technologies can amplify existing risks and create new ones, particularly in the context of AI and the metaverse.

Key Risks in the Web 3.0 Era.

1. **Aggression and Violence.**
    - **Content Risks**: Exposure to websites that encourage unhealthy or dangerous behaviors, such as self-harm, suicide, hate speech, and violent content.
    - **Contact Risks**: Experiences of harassment, stalking, and radicalization through digital platforms.
    - **Behavioral Risks**: Engagement in digital violence, humiliation of peers, and the dissemination of hateful material.

In the context of Web 3.0, AR and VR applications can create immersive environments that may exacerbate these risks. The increased involvement of children in virtual spaces heightens their exposure to cyberattacks, hacking, and threats to personal information.

Mitigating Technologies: The Internet of Things (IoT), AR/VR, blockchain, and creative artificial intelligence (AI) can either increase risks or help mitigate them by enhancing security and privacy protections.

2. **Commercial Exploitation and Management of Minors' Attention.**
    - **Content Risks**: Embedded advertising, spam, and marketing targeted at children.
    - **Contact Risks**: Tracking and collection of personal information, manipulation through AI algorithms, and misuse of data.
    - **Behavioral Risks**: Online gambling, illegal downloading, live broadcasting of abuse, and trafficking for sexual exploitation.

In the Web 3.0 era, AI and machine learning algorithms are increasingly used to engage and manipulate children's attention. Intelligent assistants and AI chatbots can influence children's thinking and behavior, potentially infringing on their rights and autonomy.

Mitigating Technologies: Artificial intelligence, AI chatbots, algorithms, and neural interfaces can either enhance these risks or be harnessed to protect children's digital rights through ethical design and regulation.

3. **Values and Worldviews Opposed to Human Rights and Sustainable Development.**
   - **Content Risks**: Idolization of chatbots, exposure to racist or discriminatory material, misleading information, and internet addiction.
   - **Contact Risks**: Ideological indoctrination and harmful beliefs that encourage self-harm.
   - **Behavioral Risks**: Creation of harmful user-generated content, incitement to discrimination, encouragement of suicide, and abuse of social media.

Web 3.0 technologies, with their greater personalization capabilities, can lead to the formation of "generative echo chambers." These environments reinforce existing beliefs and limit exposure to diverse perspectives, fostering addiction, information overload, and diminished social skills. Such echo chambers restrict children's ability to engage with a variety of ideas, impacting their decision-making and societal development.

The intersection of digital technologies and children's rights presents a complex and rapidly evolving challenge. While frameworks like the EU Kids Online classification provide a foundational understanding of digital risks, the transition to Web 3.0 introduces new dimensions that require updated approaches and robust regulatory measures. Addressing these challenges demands a proactive and collaborative effort from governments, tech companies, educators, and civil society to ensure that the digital revolution empowers the next generation rather than perpetuating inequality or infringing upon fundamental rights.

The classification developed by the European Network for Research on Children on the Internet (EU Kids Online) is relevant in relation to the challenges of Industry 3.0. EU Kids Online groups risks into three categories: content, contact and behavioural[99].

---

[99]Fourie, L. (2020). Protecting children in the digital society. In J. Grobbelaar & C. Jones (Eds.), Childhood vulnerabilities in South Africa: Some Ethical Perspectives (1st ed., pp. 229–272). African Sun Media.

1) Content risks are associated with situations where a child is a recipient of mass-produced content (through access to mass media, media and social networks).

2) Contact risks refer to situations where children actively participate in activities initiated by adults.

3) Behavioural risks are associated with situations where the child is the initiator, participant or creator of risky content or contact with peers.

Information Bubbles and Manipulation.

Social media services function as "information bubble filters" by leveraging users' digital footprints to tailor their preferences through algorithmic filtering[100]. In the face of rapid and often unregulated technological advancements, the potential for manipulating individuals becomes increasingly apparent. This manipulation operates on two levels[101]:

1. **Digital Traces**. The digital imprint left by individuals in mass communication networks and online environments.
2. **Biological and Neurobiological Data**. Insights into human biology and neurobiology allow for a deeper understanding of how individuals are structured and influenced, directly impacting decision-making patterns. This means that manipulation goes beyond merely capturing attention—it can shape personal identity.

Challenges to Human Choice.

The erosion of personal choice, evaluation, and decision-making is among the most pressing challenges facing humanity's immediate future. Increasingly, these capacities are being "outsourced" to digital platforms and technologies. However, advancements in Web 3.0—driven by startups and innovative companies—may foster positive trends by reinforcing the "right to choose." Web 3.0 aims to decentralize data control and distribute power more evenly, potentially enhancing privacy and data security. Yet, the effectiveness of these implementations remains uncertain.

---

[100]Dubois E., Blank G. The echo chamber is overstated: the moderating effect of political interest and diverse media // Information, Communication & Society. - 2018. -Vol. 21, N 5. - P. 729-745.

[101]Brett Frischmann, Evan Selinger Re-Engineering Humanity // URL:https://www.amazon.com/Re-Engineering-Humanity-Brett-Frischmann/dp/1107147093(date accessed: 26.05.2023).

Key Technologies: Balancing Risks and Benefits.

- **Artificial Intelligence (AI)**
- **AI Chatbots**
- **Algorithms**
- **Neural Interfaces**

These technologies can either amplify existing risks or mitigate them, depending on their application and regulation.

Data Privacy in the Metaverse. The advent of metaverses introduces the concept of "metaverse data" or "virtual data," where each object in augmented reality (AR) has an additional virtual layer. This raises critical questions about personal freedom, control, and security as the boundaries between the biological, physical, and digital worlds blur. The complexities of data transfer, storage, and processing in these environments necessitate enhanced security measures to prevent unauthorized access or misuse of information, especially concerning children.

Industry 4.0 and the Evolution to Web 3.0. Despite the opportunities presented by Industry 4.0 technologies, they also pose significant threats to children's rights as part of broader human rights concerns. The Internet's evolution from Web 2.0 to Web 3.0 is expected to address previous shortcomings, such as centralized data control by BigTech companies, media giants, and governments. However, Web 3.0 carries its own set of risks:

- **Increased Digital Inequality**. Limited access to new technologies and services for certain groups due to economic, geographic, or social barriers.
- **Decentralization Challenges**. While promoting autonomy and freedom, decentralization can complicate accountability and regulation, making it difficult to determine responsibility for actions within decentralized systems.

Prospects for Risk Elimination.

Eliminating these risks requires an interdisciplinary approach that bridges the gap between technological and scientific advancements and the fundamental political and legal spheres. Central to Web 3.0 should be digital identity, encapsulated in the digital rights of humans and children. Ensuring that these risks do not escalate in the era of Web 3.0 and the emerging metaverse is crucial for fostering a digital environment that supports the comprehensive development of children within the framework of sustainable development.

The Metaverse and Emerging Risks.

The expansion into the metaverse—through virtual and augmented reality technologies—introduces new layers of reality that could perpetuate existing digital vices. The creation of the dark net metaverse and the rise of AI chatbots, especially those designed without ethical norms, exacerbate these risks. These technologies can transform social relations in ways that not only exclude existing risks but also promote the full development of children as future individuals within a sustainable society.

Threats of Language Models like ChatGPT.

Language models such as ChatGPT pose significant threats when excessively used or perceived merely as tools for obtaining accurate answers rather than as creative advisors. Historian Yuval Noah Harari argues that AI threatens humanity by controlling our primary operating systems—language and speech. According to Harari, AI does not need a catastrophic event like nuclear war to dominate us; mastering language is sufficient. ChatGPT, for instance, influences our thinking and philosophy by permeating social networks, creative works, advertising, books, and scripts. These bots create virtual realities through text, making humans feel like extensions of machines without needing physical implants. Harari views ChatGPT as a tool of thought control, diminishing our ability to self-reflect and undermining what makes us human. Similarly, TikTok's algorithm curates content that keeps users engaged in an endless stream, distracting them from reality and fostering dependency.

The AI War for Individual Autonomy.

Harari describes an "AI war for each individual," where personal connections and attachments are shaped by AI in ways that are difficult to resist. He warns that we are approaching the "end of history" and the "end of the text," where people cease to engage in self-authored communication, thus diminishing their own existence. AI, in this view, represents an alien intelligence that we have unwittingly embraced, leading to a passive and entertained populace devoid of critical thought. Harari likens AI to a self-replicating weapon of mass destruction, surpassing nuclear weapons in its ability to influence and control[102].

Security Concerns from Experts.

Ross Anderson, a leading figure in security engineering, highlights that large language models (LLMs) may degrade over time,

---

[102]Lecture by Yuval Noah Harrari / URL:https://youtu.be/LWiM-LuRe6w(date accessed: 27.05.2023).

reducing the quality of the meaning they convey and thereby impacting human intelligence. This decline could weaken our cognitive abilities and decision-making processes, further entrenching AI's role in our lives.

The integration of advanced technologies in the digital age presents both remarkable opportunities and profound challenges for human and child rights. As we navigate the transition from Web 2.0 to Web 3.0 and beyond, it is imperative to address these risks through comprehensive regulatory frameworks, ethical guidelines, and inclusive governance. Only by doing so can we ensure that technological advancements serve to empower rather than undermine the fundamental rights and development of future generations.

The Real Risks and Global Responses

The risks associated with the spread of artificial intelligence (AI) are tangible, yet the question of how to regulate this growth remains unclear. Discussions at the global level revolve around possible strategies, such as setting limits on AI capabilities, imposing bans on certain types of research, or taking more drastic approaches, such as dismantling datasets altogether. However, the feasibility of enforcing international agreements on AI remains highly controversial. While global cooperation is essential, questions surrounding the enforcement of commitments from key stakeholders persist.

At a more localized level, the focus should shift toward fostering digital literacy and critical thinking. Teaching these skills is crucial in ensuring that individuals can navigate AI tools effectively and responsibly. Without this, we risk witnessing the automation of creativity rather than just routine tasks, ultimately diminishing the human ability to innovate and create.

Corporate and Geopolitical Competition

The issue of AI regulation is further complicated by fierce competition between corporations, and at the state level, between major powers like the USA and China[103]. These actors are deeply invested in the technological arms race, unwilling to cede ground in the AI market. This competition has led to the digital militarization of AI, with AI being leveraged to develop advanced weaponry, surveillance systems, and cyber warfare technologies. These developments pose a profound

---

[103]US-China Competition and Military AI / URL:https://www.cnas.org/publications/reports/us-china-competition-and-military-ai(date accessed: 27.07.2023).

challenge to global security and threaten the future of upcoming generations.

The rapid militarization of AI could lead to a new form of arms race, one that automates key areas of conflict, including:

- **Individual Capability Enhancements**: Improvements that collectively give countries a military edge.
- **AI in Decision-Making**: The growing influence of AI on national and international decision-making processes, including information management.
- **Autonomous Systems**: The rise of unmanned, AI-driven systems in warfare.
- **Intelligence, Surveillance, and Reconnaissance**: The use of AI to enhance surveillance capabilities.
- **Command, Control, and Communications**: The automation of military command structures and communications.

This stage of technological development is characterized by a race for AI supremacy, carrying significant risks for human rights, including those of children.

The Worldcoin Controversy

Another emerging concern is the biometric system linked to Worldcoin, a cryptocurrency initiative led by the CEO of OpenAI. Worldcoin seeks to establish a new form of identity verification, offering a solution to the issue of anonymous digital transactions by creating a "World ID." This protocol runs on Ethereum and aims to ensure that each digital action is tied to a real person, potentially offering new tools for developers and crypto users alike. The system also promises a universal basic income, with users receiving tokens in exchange for scanning their retinas. However, the project has raised serious privacy concerns, particularly regarding the large-scale processing of sensitive biometric data.

Countries like Germany, France, and Kenya have voiced apprehension about Worldcoin[104]. In Kenya, for example, the Ministry of Home Affairs suspended Worldcoin operations in July 2023 until government agencies could assess potential risks to public safety. Further complicating matters, a vulnerability in the Worldcoin protocol was discovered by cryptosecurity firm CertiK, which allowed attackers

---

[104]Worldcoin Digital Identity Protocol Launches with GPT-4 / URL:https://thedefiant.io/worldcoin-digital-identity(date accessed: 24.07.2023).

to bypass the retinal scan verification process, highlighting the potential security risks associated with the system.

Lack of Regulation and the Risks of Digital Transformation

The absence of effective regulation in several areas of digital transformation is another pressing issue. Among the most concerning risks are:

1. **Abuse of Data Ownership by the Private Sector**: Corporations and startups, including those collecting children's data, often exploit digital data for commercial purposes. The lack of regulation, coupled with the drive for innovation, has led to careless data handling practices, exacerbating concerns about privacy violations and the negative impacts on human rights.
2. **Digital Transformation in the Public Sector**: The transformation of outdated public sector models into "digital leviathans" presents its own set of challenges. Governments are increasingly offering "digital welfare" to citizens, but this trend risks sidelining fundamental rights and freedoms in favor of "digital security." Smart city systems, for example, may provide convenience but can also lead to surveillance overreach. Similarly, systems designed to distribute social services based on ratings, although aimed at simplifying processes, can inadvertently reduce the quality of service. AI-driven "digital defense" systems may offer enhanced state security, but they also pose existential threats to humanity if not carefully controlled.

Data Exploitation by Corporations and Startups

The issue of "data abuses by corporations and startups" primarily stems from two motivations: first, the desire to commercialize digital data to its fullest potential, and second, the ambition to enhance products by competing in the AI and advanced technology sectors, where personal user data serves as crucial raw material. When large tech companies fully control this data and use it without stringent regulations, the outcomes can be both positive and negative.

Positive Outcomes:

- **Development of the Data Economy:** The commercialization of data can drive economic growth and innovation.
- **Creation of Successful AI Systems:** Access to vast amounts of personal data can lead to the development of more effective and commercially viable artificial intelligence systems.

Negative Outcomes:

- **Loss of Privacy:** The erosion of personal privacy, especially for children, leads to a significant loss of rights and freedoms.
- **Erosion of Rights and Freedoms:** Without proper regulation, the misuse of personal data can infringe upon fundamental human rights.

To mitigate these risks, states must establish clear data ownership laws and be prepared to regulate the consequences of technological advancements effectively.

Rising Concerns Over Privacy.

Privacy has become a critical concern globally, particularly in regions where technological innovation is booming. Public trust in the tech sector has significantly declined, as demonstrated by recent events. For instance, in August 2019[105], Apple admitted that it had hired individuals to listen to audio recordings collected by its voice assistant, Siri, without user knowledge or consent. This was intended to improve Siri's machine learning capabilities for voice recognition. Such revelations highlight the urgent need for greater transparency and accountability in how companies handle personal data[106].

Digital Transformation of Public Sector Models

The digital transformation of outdated public sector models introduces new challenges, especially through the spread of "digital welfare." This trend directly impacts families and infringes upon the digital rights of children. In 2019, *The Guardian* published an investigation into the implementation of digital welfare[107], revealing the use of artificial intelligence, algorithmic modeling, forecasting, and biometric systems in managing vulnerable populations dependent on state support and social

---

[105]Patrick McGee, "Apple apologises for listening to Siri conversations," Financial Times, August 28, 2019.
3669401ba76f.

[106]ENGELKE, P. AI, Society, and Governance: An Introduction. Atlantic Council.2020.

[107]Kuntsman, A., & Miyake, E. (2022). Automated Governance: Digital Citizenship in the Age of Algorithmic Cruelty. In Paradoxes of Digital Disengagement: In Search of the Opt-Out Button (Vol. 104, pp. 41–58). University of Westminster Press. http://www.jstor.org/stable/j.ctv2z9g054.7

services. These services include child benefits, disability assistance, social housing, pensions, and employment support.[108]

Negative Impacts of Digital Welfare:

- **Automating Poverty:** Researcher Pilginton argues that the rapid digital transformation of state systems effectively "automates poverty" and "punishes the poor."
- **Exclusion Due to Digital Barriers:** Since the late 2010s, access to credit and social benefits has become exclusively available through online platforms. Many recipients lack the necessary digital skills, access to suitable devices, or even basic literacy, making it difficult to navigate these systems.
- **Complex and Opaque Processes:** The application processes are often overly complex and obscure, making it nearly impossible for recipients to understand or challenge decisions that are fundamental to their livelihoods and survival.

Lack of Transparency in Algorithmic Decision-Making.

In the digitalization of social services, as highlighted by the Child Poverty Action Group, the algorithmic decision-making process is not only inaccessible to recipients but also to support staff. Support staff do not have access to the full calculations behind benefit determinations, preventing them from explaining or altering outcomes. This lack of transparency is further compounded by frequent errors, some of which are acknowledged by the Department for Work and Pensions itself. Designing a social security system governed by algorithms represents a new form of state brutality, masquerading as efficient public service delivery. Similarly, the digitalization of policing employs technology not as a neutral tool but as a new vector for perpetuating social injustice.

Algorithmic Bias in Visa Applications

Around the world, algorithmic decision-making systems are increasingly employed to process visa applications. These algorithms often prioritize applicants based on nationality, offering "fast track" options to individuals from wealthy countries like the US, Canada, Australia, and Western Europe, as highlighted by the digital rights group

---

[108]Pilkington, Ed. 2019. 'Digital Dystopia: How Algorithms Punish the Poor'. The Guardian, 14 October. https://www.theguardian.com/technology/2019/oct/14/automating-poverty-algorithms-punish-poor

Foxglove[109]. Beyond simply allocating visas, Foxglove pointed out that these decision-making algorithms suffer from the "feedback"[110] problem—where existing biases and discrimination within the data are perpetuated and reinforced by the computer program.

A notable example occurred in 2020 when, after several years of legal challenges by Foxglove and the Joint Council for Immigrants (JCWI), the Home Office decided to scrap its biased visa algorithm[111]. This decision was widely celebrated as a victory and an acknowledgment of systemic discrimination embedded in technology design. It marked a significant milestone in the fight against biased technology. However, such incidents underscore how systemic discrimination can be ingrained in computer code and increasingly in the artificial, algorithmic technologies we rely on, often without our conscious awareness. Vulnerable groups, including children, are particularly at risk of suffering from violations of their digital rights through such biased systems.

The Need for Robust Legal Frameworks.

To prevent such abuses, it is essential to develop comprehensive legal and techno-legal frameworks[112] that restrict the unchecked activities of states and corporations in deploying digital technologies. These frameworks should also ensure equal and safe access to digital resources for everyone, including children. To prevent the benefits of digitalization from being monopolized by those who develop and implement these technologies, a new approach centered on "global digital human rights" is necessary. This includes advocating for "digital rights of the child" that are universal, regardless of geographical location.

---

[109]Foxglove. 2017. 'Legal Action to Challenge Home Office Use of Secret Algorithm to Assess Visa Applications'. Foxglove, 29 October. / URL: https://www.foxglove.org.uk/news/legal-challenge-home-office-secret-algorithm-visas (access date: 06/01/2023).

[110]Foxglove. 2020. 'Home Office Says It Will Abandon Its Racist Visa Algorithm – After We Sued Them'. Foxglove, 4 August. / URL: https://www.foxglove.org.uk /news/home-office-says-it-will-abandon-its-racist-visa-algorithm-nbsp-after-we-sued-them (date accessed: 06/01/2023).

[111]BBC News. 2020. 'Home Office Drops "Racist" Algorithm from Visa Decisions'. BBC News, 4 August. /url:https://www.bbc.co.uk/news/technology-53650758(date accessed: 01.06.2023).

[112]Farkhutdinov I. Z. International law on self-defense of states // Eurasian Law Journal. 2016. No. 1. - P. 91-100.

These principles formed the foundation of my project, **Global Digital Human Rights for the Fourth Industrial Revolution (4IR)**[113], which I implemented as a ambassador for the UN Sustainable Development Goals in Russia (2020–2021). Through this initiative and in collaboration with the Global Law Forum, I drafted a **Declaration of Global Digital Human Rights**[114], which garnered support from 50 Global Shapers centers worldwide and other youth organizations, including Future Team.

Enhancing Children's Development through Global Digital Rights

Global digital rights for children aim to create an environment that preserves their natural, barrier-free development while maximizing their potential by providing unrestricted access to the benefits of digital transformation in the Fourth Industrial Revolution.

Let us list some of the technologies of Industry 4.0, which serve as the main forces of digital transformation of society, the state and children's rights: Artificial intelligence and machine learning, Neural Interfaces, Neural network technologies, Internet of things, Big Data, modern bioengineering technologies (Biotech), Genetic Editing, Bioinformatics, Quantum Cryptography, Holography, Genome Editing, Nanorobotics, Polymer Electronics, Quantum Dots, Bionic Prosthetics, Blockchain technologies, Additive Manufacturing, Cloud computing, virtual and augmented reality (Augmented and additive reality), quantum computer (Quantum computing), cybersecurity systems (Cybersecurity), Solar Food technologies.

Identifying Problems and Barriers to Implementing Children's Digital Rights

To effectively safeguard children's digital rights, it is essential to identify the various problems and barriers that impede their implementation. These challenges can be categorized under different domains of human rights:

---

[113] Burianov M.S. Global digital human rights in the context of the implementation of sustainable development goal No. 16 // Actual problems of science and practice: Gatchina readings-2020: in 2 volumes. Vol. 2 / under the general editorship of V.R. Kovalev, T.O. Bozieva. - Gatchina: Publishing house of GIEFPT, 2020 - Vol. 2 - 487 p.-From 17-20

[114]Global Digital Human Rights for 4IR URL:http://maxlaw.tilda.ws/digitalhumanrights(date accessed: 25.05.2023).

1. **Civil Rights**
   - **Digital Divide**: Children may be excluded from access to digital technologies and online resources due to economic, geographic, or sociocultural factors. This exclusion creates significant disparities in educational and developmental opportunities.
   - **Privacy Concerns**: Children are vulnerable to privacy violations in the digital space, where their personal information may be collected, used, or disclosed without their consent or adequate protection. This undermines their right to privacy and personal security.
2. **Social Rights**
   - **Online Cruelty (Cyberbullying)**: Children may experience psychological abuse, threats, or intimidation online, which can have severe consequences for their mental health and overall well-being.
   - **Child Pornography and Exploitation**: The digital environment can facilitate the dissemination of child pornography and the sexual exploitation of children, directly violating their right to protection from such violence.
3. **Political Rights**
   - **Restrictions on Freedom of Expression**: In some countries, censorship and online control measures limit children's freedom of expression and access to information, hindering their ability to participate fully in digital discourse.
   - **Manipulation of Information, Disinformation, and Deepfakes**: Children may be exposed to misinformation, fake news, and deepfakes, which can distort their perception of reality and manipulate their opinions and beliefs.
4. **Economic Rights**
   - **Commercial Exploitation**: Children are often targets of unwanted advertising and the collection and use of their personal data for commercial purposes without their consent or sufficient transparency. This exploitation infringes upon their economic rights and autonomy.
5. **Cultural Rights**

- **Digital Disruption of Cultural and Linguistic Identity**: The pervasive influence of the Internet and digital technologies can lead to the erosion of children's unique cultural expressions and linguistic traditions.
- **Decreased Attention Function**: The prevalence of "clip-based thinking" can diminish children's ability to focus and engage deeply with information, impacting their cognitive development.
- **Restricted Access to Cultural Materials**: Children may face limitations in accessing digital resources that contain cultural materials, such as literature, music, films, and other forms of cultural expression, thereby restricting their cultural growth and appreciation.

Interconnectedness of Digital Rights Issues

It is important to note that these categories are not exhaustive, and many digital rights issues faced by children may overlap and impact multiple human rights categories simultaneously. This classification, however, helps in understanding the primary issues related to children's digital rights within the context of current digital transformation.

Growing Public Concern Over Privacy Protection

The level of public concern regarding privacy protection has naturally increased as digitalization permeates almost every aspect of human activity. Surveys indicate a declining confidence in the safety of personal data on the Internet and a decreasing belief among respondents that they possess sufficient knowledge to protect their personal data. This trend is further highlighted in the **Report on the Activities of the Commissioner for Human Rights in the Russian Federation for 2022**.

Escalating Data Breaches and Information Leaks

These concerns are exacerbated by a significant rise in data breaches and information leaks. Expert estimates suggest that in 2022 alone, such incidents increased fortyfold, resulting in the public exposure of personal information belonging to approximately 100 million individuals. Given these alarming statistics, society must prioritize strengthening information security measures and enhancing public information literacy to mitigate these risks effectively[115].

[115]Attitude of Russians to the protection of personal data. Survey of the NAFI Analytical Center in July 2021. 53 constituent entities of the Russian Federation, 1,600 respondents // NAFI Analytical Center: website. URL:

Identifying Cyber Risks in Children's Digital Development

In the study titled "Technologies for Protecting Children on the Internet," conducted in Russia from October 2021 to March 2022, researchers identified 23 cyber risks that threaten the cognitive abilities, worldview, and holistic development of children. These risks are categorized as follows:

1. **Criminalization of the Internet**
   - **Involvement of Minors in Criminal Communities**: Children being recruited into online criminal networks.
   - **Sale of Prohibited Goods and Services**: Facilitation of transactions involving illegal items.
   - **Radicalization**: Exposure to extremist ideologies and recruitment.
   - **Human Trafficking**: Use of digital platforms to exploit and traffic children.
2. **Marketing Impact**
   - **Dangerous Goods Transactions**: The Internet as a platform for purchasing items harmful to children's health and safety.
   - **Neuromarketing**: Techniques that manipulate children's neurological responses for commercial gain.
   - **Dark Patterns in User Interfaces**: Design strategies that trick children into unwanted actions.
   - **Online Fraud**: Deceptive practices targeting children for financial exploitation.
3. **Negative Impact on Child Development**
   - **Cyberbullying**: Online harassment that affects children's mental health.
   - **Stalking**: Persistent harassment or surveillance of children online.
   - **Grooming**: Manipulative tactics to exploit children sexually.
   - **Sexual Harassment**: Unwanted sexual advances or content directed at children.
4. **Exploitation in the Digital Environment**

---

https://nafi.ru/projects/it-i-telekom/otnoshenie-rossiyan-k-zashchite-personalnykh-dannykh/ (date of access: 05/25/2023).

- **Deanonymization and Doxing**: Revealing children's private information without consent.
- **Creation and Distribution of Child Pornography**: Use of digital platforms to produce and spread exploitative content.
- **Personal Data Issues**: Unauthorized collection and misuse of children's personal information.
- **Sharenting**: Parents sharing excessive or sensitive information about their children on social media.

5. **Information Pressure and Inappropriate Content**
   - **Violent Content**: Exposure to scenes of violence.
   - **Pornographic Content**: Access to sexually explicit material.
   - **Disinformation and Deepfakes**: Exposure to false information and manipulated media that distort reality.
   - **Dangerous Trends**: Participation in online challenges or trends that pose health risks.
6. **Digital Addiction and Cognitive Impact**
   - **Algorithms for Maintaining Attention**: Design features that keep children engaged for extended periods.
   - **Gaming Addiction**: Excessive gaming that interferes with daily life and development.
   - **Excessive Internet Use**: Overuse of the Internet leading to negative effects on children's cognitive functions.

Classification of Digital Rights Issues by Stakeholder and Age Group

Next, we classify the digital rights issues affecting children based on the main stakeholders in the context of Industry 4.0, segmented by age groups: up to 7 years, 7 to 14 years, and 14 to 18 years.

1. **State (Public Sector)**
   - **Children under 7**
     - **Violation of Privacy Rights**: Unauthorized collection and use of young children's personal data without parental consent and oversight.
     - **Restriction of Access to Education**: Limited access to quality and diverse online educational resources and internet availability.
   - **Children aged 7 to 14**

        - **Restrictions on Freedom of Expression**: Censorship or control over children's expression in digital environments.
        - **Online Cruelty**: Inadequate protection against cyberbullying and virtual violence, with insufficient responses to complaints and prosecution of perpetrators.
    - **Children aged 14 to 18**
        - **Restriction of Access to Information**: Limitations on access to information and content that hinder freedom of knowledge and expression.
        - **Violation of Privacy Rights**: Intrusion into adolescents' private lives through mass data collection and use without consent.
2. **Corporations (Private and Financial Sector)**
    - **Children under 7**
        - **Commercial Exploitation**: Unwanted advertising and collection of personal data from young children for commercial purposes.
        - **Insufficient Security of Online Platforms**: Risks of children accessing unsafe platforms and harmful content.
    - **Children aged 7 to 14**
        - **Data Privacy Violations**: Collection, use, and sale of children's personal data without consent or parental oversight.
        - **Behavior Manipulation and Addictions**: Use of algorithms and platform designs to attract children's attention, create addiction, and manipulate behavior.
    - **Children aged 14 to 18**
        - **Digital Inequality**: Limited access to digital resources and development opportunities due to corporations' commercial interests.
        - **Privacy and Security Issues**: Data leaks, privacy violations, and insufficient protection of teenagers' personal information in digital services.
3. **Media**

- **Children under 7**
  - **Harmful Content**: Exposure to content that incites violence, aggression, or is otherwise inappropriate for young children.
  - **Violation of Identity Rights**: Insufficient protection of children's personal identity and vulnerability on media platforms and social networks.
- **Children aged 7 to 14**
  - **Disinformation**: Spread of false or misleading information influencing children's opinions and understanding.
  - **Manipulation and Influence**: Media platforms aiming to manipulate and influence children's behavior and opinions.
- **Children aged 14 to 18**
  - **Psychological Pressure and Impact on Self-Esteem**: Constant comparisons with others and idealization of life through social media, leading to psychological pressure and low self-esteem.

4. **Family as Part of the World Community**
   - **Children under 7**
     - **Lack of Control and Supervision**: Limited parental control over young children's online activities and insufficient digital safety training.
     - **Risk of Contact with Unwanted Individuals**: Possibility of unintentional contact with potentially dangerous individuals online.
   - **Children aged 7 to 14**
     - **Information Overload**: Excessive information and content without proper guidance and support.
     - **Inappropriate Content**: Risk of accessing content not suitable for their age and development.
   - **Children aged 14 to 18**
     - **Restriction of Freedom and Control**: Family control and restrictions over teenagers' digital activities limiting their freedom of expression and autonomy.

- **Lack of Education about Digital Risks**: Insufficient training for parents on digital risks and lack of active participation in teenagers' digital lives.

5. **Educational Institutions**
   - **Children under 7**
     - **Limited Access to Digital Education**: Educational institutions' limited capacity to provide quality and accessible digital education for young children.
     - **Lack of Digital Safety Training**: Insufficient inclusion of digital safety, digital literacy, and awareness in education programs for children.
   - **Children aged 7 to 14**
     - **Digital Divide in Education**: Limited access to digital resources and technologies within educational institutions, exacerbating the digital divide among children.
     - **Insufficient Critical Thinking Training**: Inadequate training in critical thinking and digital literacy to effectively navigate the digital information space.
   - **Children aged 14 to 18**
     - **Lack of Support in Choosing Careers in Technology**: Insufficient guidance and support for teenagers in choosing and developing careers in digital technology, startups, and projects to overcome the digital divide.

Future Prospects and Mitigation Strategies

With the development of the Web 3.0 economy, many individuals will gain increased control and choice over the personal data they share with online systems and corporations. The concept of decentralized identity may play a significant role in this shift, although its implications for minors remain unclear. Additionally, generative AI will provide greater access to personal teachers, development tools, and participation in the digital revolution, potentially minimizing the risks outlined in the contexts of educational institutions and families.

One significant area of challenge and opportunity is neurotechnologies, which are transitioning from experimental projects to real initiatives and startups. As UNESCO notes, designing and implementing well-thought-out and effective norms and principles to ensure that neurotechnologies are developed and deployed ethically, in the interests of individuals and society, requires careful identification and characterization of the issues at hand. This involves:

- **Understanding the Technologies Being Developed**: Identifying what neurotechnologies are being created.
- **Identifying Developers and Their Locations**: Knowing who is developing these technologies and where.
- **Comprehending Interactions with Other Technological Trends**: Understanding how neurotechnologies interact with other advancements, especially AI.

Defining Key Challenges and Structuring Solutions.

In conclusion, we identify the main challenges that create barriers to the implementation of children's digital rights across both Industry 3.0 and Industry 4.0 eras, which include the internal revolutions of generative AI and other advanced technologies. These risks can negatively affect children's physical, psychological, and social safety, impacting their attention, memory, physical development, and rights to privacy, information, and freedom of expression.

Complex Cyber Risks to Children's Rights in Industry 3.0.
During the third industrial revolution, the integration of the Internet and smartphones introduced a multitude of cyber risks affecting children's rights. These risks encompass:

- **Cybercrime**: Online fraud, cyberbullying, and child pornography.
- **Cyber Risks**: Exposure to inappropriate content, privacy violations, and identity theft.
- **Internet Governance Issues**: Challenges related to data protection, digital rights, and access to information.

These risks can severely impact children's physical, psychological, and social safety, affecting their attention, memory, physical development, and their rights to privacy, information, and freedom of expression.

Emerging Risks in the Fourth Industrial Revolution.
As we transition into the Fourth Industrial Revolution, characterized by Web 3.0, metaverses, and generative artificial intelligence, new threats and challenges emerge:

- **Privacy and Data Security**: Enhanced risks related to the protection of personal data.

- **Mental Health Impacts**: Negative effects on attention, memory, and cognitive functions, as well as participation in the physical world.
- **Accessibility and Quality of Education**: Issues surrounding equitable access to digital educational resources.
- **Digital Control and Uncertainty**: Increased control mechanisms that may unpredictably impact children's development and rights.
- **Protection from Exploitation and Violence**: Enhanced threats to children's safety and well-being in digital environments.

Syncretic Digital Risks.

A significant concern arises when digital risks reinforce each other, creating syncretic effects—integral combinations of heterogeneous elements that amplify overall threat levels. We introduce the term **"syncretic digital risks"** to describe these compounded threats. These risks emerge from the fusion of the digital (virtual) and biological worlds and are driven by the exceptional potential of technologies without corresponding ethical, public-legal, educational, and political safeguards.

Syncretic digital risks represent the most significant threats of the upcoming digital transformation, arising at the intersection of Fourth Industrial Revolution technologies, Web 3.0, and knowledge-intensive innovations. These risks are caused by the rapid advancement in scientific and technological fields and the lag in ethical, public-legal, educational, and political spheres. This imbalance leads to an exponential digital transformation of outdated social relation models.

As part of the theoretical study of future digital risks, we propose the concept of **"generative panopticon echo chambers"**. This term characterizes a worst-case scenario for a digitally transformed society, where:

- **Digital Surveillance**: Extensive monitoring of individuals.
- **Sovereignty of Internet States**: Each state controlling its own internet environment.
- **Manipulation through Language Models**: Use of sovereign or malicious AI models (such as versions of GPT) to shape individual thought processes.
- **Total Surveillance**: Complete oversight of societal members, reducing creativity and critical thinking abilities.

Cognitive Digital Risks for the Individual. Within the framework of **"cognitive digital risks for the individual"** in the 21st century, we introduce the concept of **"digital desubjectification"**. This multi-level process involves the regression of an individual's fundamental foundations and essential capabilities due to the rapid introduction of digital technologies that minimize the right to freedom of choice and decision-making. This leads to neurobiological consequences such as:

- **Dependence on Technology;**
- **Objectification by Technology;**
- **Person as a Servant of Technology.**

These risks stem from two primary sources:

1. **Digital Trace or Imprint**. The pervasive presence of an individual's digital footprint in mass communication networks and online environments.
2. **Biological and Neurobiological Data**. The utilization of biological and neurobiological insights to influence and manipulate decision-making patterns.

Stages of Digital Desubjectification

1. **Digital Social Mining**. Extracting valuable information from databases, social networks, and other digital platforms to analyze user behavior and preferences. While this allows organizations to make informed decisions, it also exploits basic human needs by creating attachments that reduce the ability to choose freely. For instance, data analysts can identify public opinion trends or predict consumer behavior, making minors particularly vulnerable as platforms are often tailored to adult audiences. It is crucial to prevent the collection of data from individuals under 18.
2. **Digital Social Programming**. Designing systems that manipulate human behaviors and thoughts, transforming individuals into subjects controlled by digital systems rather than autonomous beings.

The Role of Programmers in Digital Legislation. Today's challenges in digital transformation fully justify the adage, **"programmers are the unrecognized legislators of the world."** The future digital society, whether it leads to digital prosperity or poverty, will depend on the values embedded in digital products—from the initial ideas of startups to the operations of large technology corporations. These risks are associated with the darker aspects of the digital world, including existential threats posed by general artificial intelligence (AGI) and other technologies that extend beyond direct human and child rights issues. We define these as the most significant challenges threatening modern human civilization, encompassing:

- **Violations of Critical Information Infrastructure;**
- **Failures in Technology Management;**
- **Non-Compliance with Global Cybersecurity Measures;**
- **Digital Concentration of Energy;**
- **Man-Made Environmental Damage;**
- **Geopoliticization of Strategic Resources;**
- **Dominance of AGI over Human Interests.**

Structured Approach to Addressing Challenges

For a systematic approach to analyzing the problems of implementing children's digital rights, it is essential to structure the challenges by levels:

1. **Scientific-Theoretical Level**: Understanding the underlying technological and cognitive impacts on children.
2. **Normative-Legal Level**: Developing and enforcing regulations that protect children's digital rights.

Law Enforcement Level: Ensuring compliance and addressing violations through appropriate legal mechanisms)[116].

For a systematic approach to the analysis of problems of implementing children's digital rights, it is necessary to structure the problems by levels - scientific-theoretical, normative-legal and law enforcement - and propose solutions aimed at overcoming them.

1. Scientific and theoretical level

At this level, the problems are related to the lack of a clear definition of the concepts and concepts necessary for a full understanding and implementation of the digital rights of the child:

1. **Lack of a clear definition of children's digital rights and related concepts**: There is no universally accepted definition of children's digital rights, which includes not only rights to protection, but

---

[116]Burianov M.S. Empowerment in social work: theory and technology of empowerment: a tutorial / T.S. Kienko, P. Ya. Tsitkilov, L.A. Kaigorodova, S.G. Furdey, L.S. Detochenko, S.A. Buryanov, O.M. Papa, E.A. Blagorodova, M.S. Konstantinov, E.I. Shashlova, E.V. Martynova, D.V. Davydenko, M.S. Buryanov, I.N. Gnedysheva (under the general editorship of T.S. Kienko); Southern Federal University. - Rostov-on-Don; Taganrog: Publishing House of the Southern Federal University, 2021. - 255 p. P. 126-153.

also rights to access, choice and safety in the digital environment and use of technologies.

2. **Fragmented approaches to digital rights**:

- **Limited by the human rights perspective**: The focus is predominantly on protecting children in the digital environment, while the rights of the child should include a wide range of rights in the context of digital transformation (access to the Internet, information rights, the right to development, etc.).
- **Narrow focus on information rights**: Children's digital rights are often considered solely in the context of Industry 3.0 (the Internet and smartphones), without taking into account the possibilities of Industry 4.0 (generative AI, Web 3.0 and other technologies).

3. **Insufficient attention to universal child rights in the context of digital transformation**: The transformation of children's rights in the digital age requires a revision and expansion of the legal framework to include new digital rights (e.g. the right to be forgotten).

4. **Failure to take into account all stakeholders**: Insufficient attention is paid to the role of all actors (government bodies, the private sector, educational institutions, media and citizens) in realizing the digital rights of the child.

5. **Lack of integration of the concepts of "digital identity" and "digital citizenship"**: Insufficient development of concepts that reflect the new challenges and opportunities of digital transformation.

2. Normative and legal level

At this level, the problems are related to the lack of adequate legal acts and international treaties that would regulate the digital rights of children in the context of modern digital transformation:

1. **Lack of legal regulation of key digital challenges**:

- **Digital Militarization**: Threat to future generations due to lack of legal control over the use of military technologies in the digital space.
- **Digital Divide**: Problems of children's access to innovative technologies (artificial intelligence, quantum technologies, etc.), which are not regulated by international law.
- **Digital Total Control and Privacy**: Insufficient protection of children's data and privacy in the digital environment.
- **Risks of Artificial Intelligence and Algorithms**: Uncertain legal framework for the governance of AI and machine learning that impact children's rights.

2. **Insufficient regulation of risks of industries 3.0 and 4.0**:

o **Cyber risks in the context of the third industrial revolution**: Complex cyber threats (e.g. cyberbullying, data leaks) remain insufficiently regulated at the national level.

o **Risks of the Fourth Industrial Revolution**: Uncertain legal frameworks for new technologies (Web 3.0, metaverses, generative AI) and their impact on children's rights.

3. **Lack of enshrinement in international law**:

o International treaties and principles do not include the concept of children's digital rights in the context of Industry 4.0. General Comment No. 25 (2021) on children's rights in the digital environment is not legally binding.

3. Law enforcement level

At this level, the problems concern the insufficient adaptation of law enforcement practices and mechanisms to the realities of the digital age:

1. **Lack of a human-centered approach**: Current law enforcement practices do not take into account the unique characteristics of children's interaction with the digital environment and require revision to take into account their specific needs and rights.

2. **Lack of public oversight and parental involvement**: Lack of mechanisms to ensure parental and public participation in decision-making regarding children's digital rights.

3. **Problems in the activities of international organizations**:

o **Duplication of functions**: Individual agencies and organizations duplicate programs, leading to inefficiencies.

o **Predominance of declarative approaches**: A tendency towards self-regulation and ethical norms instead of clear legal obligations, which reduces the effectiveness of the implementation of children's rights.

4. **Insufficient training of personnel**: At the national level, for example in Switzerland, there is a lack of training for child protection professionals in the context of digital transformation, which limits their ability to effectively address new challenges.

To solve the identified problems, a comprehensive approach is required, including:

- Development of clear scientific and theoretical foundations and concepts in the field of children's digital rights.

- Creation of legal acts governing the digital rights of children at the national and international levels, taking into account Industry 4.0 technologies.
- Implementing human-centered law enforcement practices that take into account the unique needs of children in the digital age.
- Strengthening international cooperation and coordination to eliminate duplication of functions and ensure the effective implementation of children's digital rights.

3.2. Prospects for improving legal norms and mechanisms for implementing digital rights and the world

Let us navigate cyberspace together, where every movement, emotion, word, and desire becomes another byte in a vast sea of data. In a world where this data is not only recorded but also manipulated and used to steer us through intricate incentive matrices for someone else's profit and control, it is time to find a way to a human-centric digital world.

The journey begins with exploring the prospects of creating an innovative regulatory climate that ensures digital transformation serves the interests of childhood. This involves expanding children's capabilities in the digital realm without negatively impacting their cognitive functions and health. Today, digital technologies have become a new environment, a novel mode of interaction, and a fresh source of knowledge. Consequently, children's rights are being transformed and adapted to this new reality—they are becoming digital, global, and integral to our lives. For instance, in Russia, 53% of children spend one to four hours a day on digital devices, while more than a quarter (26%) spend all their free time on them[117].

In this context, the digital rights of children are evolving from a declarative concept into a tangible necessity. These rights serve as tools to create order in the digital world, protecting children from the abuses of omnipresent corporations and the overreach of non-adaptive states.

---

[117]Digital Habits: Every Fourth Child Spends All Their Free Time on Gadgets URL:https://www.kaspersky.ru/about/press-releases/2021_cifrovye-privychki-kazhdyj-chetvyortyj-rebyonok-provodit-v-gadzhetah-vsyo-svobodnoe-vremya(date accessed: 26.07.2023).

Establishing clear boundaries and rules is essential to harnessing children's potential, providing them with freedom and opportunities for development. This involves not only implementing protective measures based on restrictions—such as UNESCO's proposed global ban on smartphones in schools, highlighted in the July 2023 report—but also creating conditions that foster children's development through tailored access to digital technologies.

In resolution A/RES/78/265 it was stated for the first time that special attention must be given to those in vulnerable situations. Yet for now this remains only a declaration. In my concept of digital rights, I develop the idea of special immunities for children and minorities: a prohibition on the exploitation of data, mandatory filters for harmful content, and technological "safety zones" embedded into digital platforms — or technologies and platforms designed by default to be age-appropriate. In this regard, let us explore the prospects of using the latest technologies to uphold children's rights.

Biometric Identification Technologies.

With the rapid introduction of new technologies like Web 3.0, it is essential to develop a legislative framework for the right to digital identity. This right will be a cornerstone of digital human rights, ensuring accessibility and the ability to benefit from Fourth Industrial Revolution technologies, including essential services like education, healthcare, and social security. Additionally, the right to digital identity can form the basis for systems providing basic income in various forms. However, to effectively implement these rights globally and ensure they are accessible to everyone, attention must be paid to protecting digital identities from malicious actions such as hacking, fraud, and illegal data collection. This requires the development and adoption of international security and privacy standards.

For example, Worldcoin, a startup developed by Sam Altman that scans irises, is powered by generative AI. To ensure that such initiatives benefit all of humanity, it is necessary to legally enshrine global digital rights for individuals and children. Concurrently, the younger generation must be educated and equipped with tools for digital literacy and principles for interacting with decentralized platforms amidst emerging technologies.

Genetic Editing.

Genetic editing holds the potential to treat hereditary diseases in children, thereby promoting their right to a healthy and fulfilling existence. It can also enhance the effectiveness of medical drugs by tailoring them to individual needs, further supporting children's health rights in a digital context.

Bioinformatics.

Bioinformatics can be leveraged to ensure the security and privacy of children's medical data. This innovative field contributes to the development of reliable information storage systems and robust encryption algorithms, thereby protecting the privacy of children's health and personal information.

Neurointerfaces.

Neurointerfaces enable interaction with digital devices using thoughts or brain signals. In the context of children's digital rights, this advanced technology can help children with disabilities participate in educational processes and communication, providing new means of self-expression and ensuring their active involvement in digital society.

Quantum Cryptography.

Quantum cryptography offers advanced encryption methods to secure digital communications and protect children's data online. This technology can prevent unauthorized access to personal information, creating a safer online environment for children.

Holography.

Holographic technologies in virtual and augmented reality can provide children with new educational and entertainment opportunities while ensuring their safety and respect for privacy. These innovations support the development of educational applications, virtual tours, and interactive tools that enhance knowledge acquisition and talent development.

Augmented Reality (AR).

AR helps develop various skills in children, including research abilities, spatial awareness, and problem-solving. By creating hybrid digital learning spaces that blend digital and physical objects, AR fosters communication and critical thinking skills[118].

Navigating the Metaverse Towards a Human-Centric Digital World

Within the concept of the metaverse, technologies hold immense potential when the virtual and real worlds merge in a way that preserves authentic social relationships. According to the UNICEF Metaverse, Extended Reality, and Children Report, it is crucial to maintain genuine

---

[118]Nekoui, Y., & Roig, E. Playgrounds in the digitally mediated city: An approach from augmented reality. Urbani Izziv, 33(2). 2022. – p.82–90.

interactions between teachers and children, caregivers and children, and among peers. In this digital transformation, the key digital rights of children will be the right to education from anywhere in the world and the right to access global digital technologies.

However, maximizing the potential of digital technologies is impossible without systematic efforts focused on legal regulation and the protection of digital rights. Currently, legal regulation of new technologies primarily addresses private law aspects, while public law regulation remains largely overlooked. In some countries, including Russia, there are emerging attempts to ethically regulate aspects of digitalization, but significant gaps still exist.

Informational Rights in the Context of Industry 3.0 and Beyond.

Most researchers initially emphasize informational rights in the context of Industry 3.0 rather than modern technologies. These rights include:

- **The Right to Information**: The ability to seek, receive, transmit, and distribute information.
- **The Right to Protect Data and Information**: Ensuring confidentiality and safeguarding personal data.
- **The Right to Protection from Information Impact**: Protecting individuals from the adverse effects of information overload and manipulation, including in mental and neuro-related spheres in some countries.

Controversies in Russia: Biometrics and Facial Recognition in Educational Institutions.

In Russia, one of the most contentious issues is the implementation of children's digital rights concerning biometrics and facial recognition systems. Current regulations for installing video surveillance in educational institutions for anti-terrorist security purposes do not prohibit the use of facial recognition technology. According to the Government of the Russian Federation's Decree No. 1006 (August 2, 2019), educational institutions must ensure access control for visitors and store video surveillance recordings for one month. This requirement renders the use of facial recognition in these systems potentially unnecessary and inappropriate.

The Ministry of Digital Development, Communications and Mass Media of Russia, the Ministry of Education of Russia, and Roskomnadzor expressed their stance against using facial recognition in educational institutions during a working meeting in April 2021, organized by the

Commissioner for Children's Rights under the President of the Russian Federation. Despite this, as of January 2020, Roskomnadzor reported that biometric systems were implemented in educational organizations across at least 18 constituent entities of the Russian Federation.

Under Article 19 of the Federal Law No. 152-FZ "On Personal Data" (July 27, 2006), educational institutions acting as data operators must implement organizational and technical measures to secure processed personal data. Financial constraints preventing educational institutions from ensuring physical security further exacerbate concerns about the security of students' personal data.

In response, the Children's Rights Commissioner proposed amending the relevant government resolution to prohibit the installation of cameras with facial recognition technology and the processing of students' biometric data in educational institutions. The Russian Ministry of Education, representing the Russian Government, received support for this ban from the Russian FSB and the Ministry of Digital Development. However, the Ministry of Internal Affairs and the Russian National Guard opposed the ban, arguing that their proposed city security system does not require the collection of biometric data from minors, as it relies on comparing images with photographs provided by law enforcement agencies.

Increasing Digital Risks with Technological Advancements.

With the evolution of the Internet from Web 2.0 to Web 3.0, alongside advancements in big data, artificial intelligence, and neural networks, society's digitalization is accelerating, leading to heightened digital risks. The uncontrolled use of not only general information but also personal data is becoming increasingly prevalent. Even "anonymous" personal information is at risk, as data from different sources can be linked through common fields in data sets, especially in the context of data and database leaks.

Ethical Standards and the Role of Technology Companies.

In October 2021, following the event "Ethics of Artificial Intelligence: The Beginning of Trust," Russia's leading technology companies agreed on a declarative Code of Ethics in the Sphere of Artificial Intelligence. This code emphasizes the development of artificial intelligence based on principles of non-discrimination, data security, information security, respect for human autonomy, and responsibility for the consequences of AI use.

A Paradigm Shift Towards Comprehensive Legal Frameworks.

In response to the myriad risks of the digital era, a fundamental paradigm shift is essential. Ethical approaches and corporate self-regulation alone are insufficient for effectively implementing the digital rights of humans and children. The cornerstone of digitalization must be the legal consolidation of the latest generation of human rights and digital rights specifically tailored for children. To achieve this, we introduce the concept of **"global digital rights,"** which encompasses[119]:

1. **Access to Global Digital Technologies**: Ensuring every child can utilize digital tools and technologies worldwide.
2. **Protection of Personal Information**: Safeguarding personal data in the context of digital technology use.
3. **Expansion of Human Capabilities**: Leveraging digital technologies to enhance human potential while upholding all forms of human rights.

Defining Basic Digital Human Rights[120]:

1. **Right to Access the Global Internet**: Including digital individualization, storage of digital assets, and other digital technologies.
2. **Right to Access Reliable Information**: Ensuring children can obtain accurate and trustworthy information.
3. **Right to Privacy and Data Protection**: Protecting personal, genetic, and biometric data.
4. **Access to Social Services via Techno-Legal Platforms**: Utilizing digital platforms to provide essential social services.
5. **Digital Access to Education and Cultural Values**: Ensuring equitable access to digital education and cultural resources.
6. **Right to Participate in Property Circulation**: Enabling children to engage in digital transactions and property exchanges.

---

[119]Burianov M.S. Digitalization of Law in the Context of Globalization // Globalization and Public Law: Proceedings of the International Scientific and Practical Conference of November 22, 2019 - Moscow: RUDN University, 2020. - P. 91.

[120]Burianov M.S. Gateway to Global Law: Global Digital Human Rights // Scientific Works of the Moscow Humanitarian University. / M.S. Buryanov. - 2020. No. 2. - P. 63-66.

7. **Priority of Human Rights in Technology Use**: Emphasizing the use of technologies to uphold human rights, including prohibiting technologies that harm humans, particularly artificial intelligence.

Enshrining **global digital human rights** paves the way for creating international laws that support sustainable development.

Establishing a Global Digital Regulatory System.

The rapid pace of digitalization necessitates the formation of a **global digital regulatory system**. This system should consist of principles and norms encoded in software, machine programs, and mathematical solutions, particularly focusing on cryptography to protect human rights. For children's rights, implementation measures must be preventive and effective, tailored to different age groups. These measures should aim to maximize freedoms by enhancing digital literacy and adapting digital platforms to suit children's needs without imposing strict content restrictions.

Integrating Digital Rights into Modern Technologies.

It is crucial to embed digital rights for individuals and children directly into the algorithms of modern technologies and techno-legal principles. This integration is vital to mitigate risks associated with cutting-edge technologies such as Web 3.0 and generative artificial intelligence (AI), including their combinations like the Worldcoin cryptocurrency, which raised $115 million by mid-July 2023.

Approaches to AI Implementation:

1. **Accelerated Implementation**: Some regions advocate for the rapid adoption of AI across various sectors, especially those impacting society broadly. For example, Yokosuka City in Japan was the first in the world to hire ChatGPT, conducting a month-long experiment that demonstrated increased work efficiency and reduced working hours for city administration employees. Notably, Japan initially considered allowing generative AI to bypass copyright laws but later retracted this idea.
2. **Total Restriction and Bans**: Conversely, other countries adopt a more restrictive stance. Italy, for instance, banned the use of ChatGPT, setting a precedent for European regulation. Both approaches are quite radical, highlighting the need for a balanced middle path that incorporates modern legal regulations while enhancing access to advanced technologies for humanity.

Defining the Key Digital Rights of the Child by Age Group. To ensure the effective protection of children's digital rights, it is essential to categorize these rights based on age groups: children under 7 years old; children from 7 to 14 years old; and children from 14 to 18 years old. Below, we outline the key digital rights for each age group within these classifications:

1. **Right to Privacy and Data Protection**
   - **Children Under 7 Years Old**: Prohibition on processing and using the child's personal data.
   - **Children from 7 to 14 Years Old**: Prohibition of processing personal data without the consent of parents or legal guardians, and ensuring control over the collection and use of such data.
   - **Children from 14 to 18 Years Old**: The right to self-determination and the ability to delete personal data.
2. **Right to Equal Access and Participation in the Global Internet**
   - **Children Under 7 Years Old**: The right to safe and educational access to digital resources and age-appropriate content.
   - **Children from 7 to 14 Years Old**: The right to information literacy and protection from harmful and inappropriate content.
   - **Children from 14 to 18 Years Old**: The right to freedom of expression and participation in the digital sphere, including access to digital personalization tools.
3. **Right to Education and Development**
   - **Children Under 7 Years Old**: The right to access quality digital education and development technologies.
   - **Children from 7 to 14 Years Old**: The right to develop digital skills and critical thinking within the digital environment.
   - **Children from 14 to 18 Years Old**: The right to career guidance and access to digital resources for future career development.
4. **Right to Security and Protection**
   - **Children Under 7 Years Old**: Protection from unwanted contacts and exploitation in the digital environment.

- **Children from 7 to 14 Years Old**: Protection from cyberbullying, online harassment, and harmful influences.
- **Children from 14 to 18 Years Old**: Protection from digital discrimination and online violence.

5. **Right to Equality and Inclusion**
   - **Children Under 7 Years Old**: Ensuring equal opportunities for access to digital resources and technologies.
   - **Children from 7 to 14 Years Old**: Preventing discrimination in digital environments and access to digital applications.
   - **Children from 14 to 18 Years Old**: Protection from digital discrimination based on gender, race, nationality, or other categories.
6. **Right to Ethical and Legal Use and Development of Technologies**
   - **Children Under 7 Years Old**: Ensuring the safety and ethical standards of digital products and services intended for children, with content adapted rather than strictly regulated.
   - **Children from 7 to 14 Years Old**: Supporting the development of legal norms and values in the use of digital technologies.
   - **Children from 14 to 18 Years Old**: Promoting children's participation in the development of legal and responsible digital technologies.
7. **Right to Protection from Digital Inequality and Discrimination**
   - **Children Under 7 Years Old**: Preventing digital and information inequality in access to resources and opportunities.
   - **Children from 7 to 14 Years Old**: Combating the digital divide and ensuring equal access to educational and development technologies.
   - **Children from 14 to 18 Years Old**: Protection from digital discrimination based on gender, race, nationality, or other categories.
8. **Right to Transparency and Control Over the Use of Data (Personal, Genetic, and Biometric Data)**

- **Children Under 7 Years Old**: Guaranteeing data protection and confidentiality of personal, genetic, and biometric data.
- **Children from 7 to 14 Years Old**: Ensuring transparency in the collection, storage, and use of biometric data.
- **Children from 14 to 18 Years Old**: The right to control and participate in decisions regarding the use of their personal, genetic, and biometric data.

9. **Right to Legal Use and Development of Artificial Intelligence**
   - **Children Under 7 Years Old**: Protection from the harmful and inappropriate use of artificial intelligence concerning children.
   - **Children from 7 to 14 Years Old**: Ensuring transparency and understandability of algorithms and AI-based solutions.
   - **Children from 14 to 18 Years Old**: Participation in debates and decision-making on ethical issues related to AI use. The right to access generative AI technologies in the context of "generative knowledge" must be accompanied by a high level of critical thinking and understanding of the technology.
10. **Right to Know Your Digital Rights and Freedoms in an Accessible Form**
   - **Children Under 7 Years Old**: The right to understand their rights and freedoms in a format adapted to their age, facilitated through preschool organizations to provide digital hygiene skills.
   - **Children from 7 to 14 Years Old**: The right to be informed about their digital rights and freedoms in a manner appropriate for their school age, including educational programs and interactive learning materials to develop safe and responsible digital technology usage.
   - **Children from 14 to 18 Years Old**: The right to access more detailed and in-depth information on digital rights and freedoms appropriate to their age and understanding level, through educational courses, workshops, and events aimed at preparing them for adult life in a digital society, including digital ethics and responsibilities.

11. **The Right to Informed Consent and Data Protection in the Context of Neural Interfaces, Neurotechnologies, Neuromarketing, and Systems Using Big Data**
    - **Children Under 7 Years Old**: Guaranteeing the protection of personal data collected through neurotechnology, emphasizing the need for parental or guardian consent, along with a full explanation of the potential use of such data.
    - **Children from 7 to 14 Years Old**: Providing the opportunity for informed consent regarding the use of neurotechnology and neurointerfaces, with appropriate explanations about the collection and purpose of brain activity data, and detailed information on data protection measures.
    - **Children from 14 to 18 Years Old**: Allowing active participation in decision-making processes related to the collection and use of brain activity data, ensuring full understanding of consent terms and confidentiality, and providing tools to control and manage their data in accordance with applicable laws and ethical standards. Referencing BBC News, successful experiments in implanting devices to improve mobility for paralyzed individuals—where implants transmit signals to sensors on a helmet—highlight the urgent need to develop new rights in this area now.
12. **The Right to Forget Data That Is Not Critical to a Person's Further Development**
    - **Children Under 7 Years Old**: Not directly applicable as this right is more relevant upon reaching adulthood.
    - **Children from 7 to 14 Years Old**: Not directly applicable as this right is more relevant upon reaching adulthood.
    - **Children from 14 to 18 Years Old**: The right to forget non-critical data (such as certain medical data that must be stored in a single register) upon reaching adulthood. With the proliferation of digital spaces and Internet of Things technologies collecting and profiling data, concerns arise about digital dossiers that can accompany young people into adulthood, affecting their access to education, employment, healthcare, and financial

services. Thus, every individual who reaches adulthood should be guaranteed the right to erase their digital traces and data accumulated during childhood to minimize risks related to digital profiling and digital dossiers.

Formulating Principles for Implementing Children's Digital Rights

To effectively protect and promote the digital rights of children, we must establish clear principles tailored to different age groups: children under 7 years old; children from 7 to 14 years old; and children from 14 to 18 years old. Below, we outline these key principles:

1. **The Principle of Equal Access to Digital Rights.**
   - **Ensuring Equal Opportunities**: Guaranteeing all children access to Industry 4.0 digital technologies, tailored to their specific age groups (under 7, 7-14, 14-18). Instead of imposing complete restrictions, platforms should be adapted to meet the needs of each age group, thereby overcoming the digital divide.
2. **The Principle of Non-Alienation of Children's Digital Rights and Freedoms.**
   - **Integral Rights**: Recognizing that children's digital rights and freedoms are an integral and indivisible part of their universal rights and cannot be deprived of them.
3. **The Principle of Active Participation of the Child in the Digital Environment.**
   - **Inclusion in Development**: Incorporating children's opinions and views in the development and use of digital technologies. Ensuring their participation in decision-making processes that affect their digital rights and freedoms.
4. **The Principle of the Best Interests of the Child in the Digital Environment.**
   - **Priority to Well-Being**: Prioritizing the interests and well-being of children in the development and use of digital technologies. This includes reasonable protection from harmful content, cyberbullying, privacy violations, and other cyber risks.

5. **The Principle of Digital Pluralism and Opportunities for Child Development.**
    - **Diverse Perspectives**: Creating conditions that allow for a diversity of viewpoints in implementing children's digital rights, preventing the ideologization of the digital environment. Emphasizing digital literacy and skills development rather than restricting internet access to foster a conscious and responsible attitude toward global digital technologies.
6. **The Principle of Special Care, Assistance, Safeguarding, and Protection for All Children in the Digital Environment.**
    - **State and Societal Responsibility**: Mandating that the state and society provide special care, assistance, and protection to all children, especially those in vulnerable situations. This includes providing appropriate tools, resources, and support to develop digital literacy and protect against potential threats.
7. **The Principle of Long-Term Perspective in Updating Legal Regulation of Children's Digital Rights.**
    - **Future-Proof Rights**: Recognizing that while universal human rights have remained relatively unchanged over centuries, the rapid technological advancements of the third, fourth, and upcoming fifth industrial revolutions require rights that anticipate future technological developments.
8. **The Principle of Digital Rights in Digital Platforms by Default.**
    - **Built-In Protections**: Ensuring that digital platforms inherently support the implementation of children's digital rights. This involves integrating protective measures within digital systems from the outset.

Expanding Digital Rights for the Future Web 3.0 Society.

In the context of the final principle, we present potential digital rights for the future child, which will form the foundation of a Web 3.0 society, encompassing new startup cities, network states, and decentralized global institutions:

- **The Right to Digital Sovereignty**
    - **Control Over Digital Identity**: Every child has the right to control their digital identity and data, determining

what information is collected, how it is used, and with whom it is shared. This ensures digital autonomy and protection from potential abuse.

- **The Right to Quality Global Education in the Digital Age**
    - **Access to Education**: Every child has the right to access quality education anywhere in the world, ensuring understanding of digital technologies, their ethical and legal use, and the development of systemic, critical thinking and creative potential.
- **The Right to Digital Consciousness and Digital Autonomy**
    - **Informed Participation**: Every child has the right to conscious participation and control in the digital environment. This includes ensuring that decisions made by automated AI-based systems or quantum computers are transparent, explainable, and predictable, respecting individual values and preferences.
- **The Right to Digital Responsibility and Ethics**
    - **Ethical Technology Use**: States, technology companies, and society as a whole are responsible for developing, implementing, and using advanced technologies ethically. This involves assessing potential negative consequences for people and the planet and preventing abuses and human rights violations in the digital sphere.
- **The Right to Digital Security and Protection from Existential Threats**
    - **Protection from Severe Digital Risks**: Every child has the right to be protected from digital threats that may have existential consequences for humanity. This requires robust security mechanisms to prevent data manipulation, cyber-attacks, and the misuse of advanced technologies that could threaten human life and well-being.
- **The Right to Startup Development**
    - **Support for Innovation**: Every child has the right to support their innovative initiatives. This should be reflected in the startup ecosystems of countries, providing access to funding and infrastructure to test and develop ideas.

Implementing Children's Digital Rights: Key Steps

To effectively implement these digital rights, the following steps should be taken:

1. **Digital Literacy Training**
   - **Educational Programs**: Provide training programs and resources to help children develop digital literacy skills, including awareness of safe and ethical behavior in the digital environment.
2. **User Settings and Controls**
   - **Parental Empowerment**: Empower parents and guardians to set limits and controls on AI chats and similar services to ensure their use is appropriate for the child's age and needs.
3. **Partnerships with Child Protection Organizations**
   - **Collaborative Efforts**: Collaborate with international and national organizations specializing in child protection in the digital environment to share knowledge, experiences, and best practices.
4. **Transparency and Feedback**
   - **Operational Clarity**: Ensure transparency in the operation of AI chat technologies and policies used by children, and incorporate user feedback from parents and children to continuously improve and meet their needs.
5. **Research and Development**
   - **Innovative Approaches**: Conduct scientific research and develop innovative approaches to ensure a safe, inclusive, and nurturing digital environment for children.
6. **Promoting Meaningful Inclusion**
   - **Diverse Representation**: Promote the meaningful inclusion of all children, especially girls, children from disadvantaged backgrounds, children with disabilities, and underrepresented children, ensuring their voices are heard and respected in the digital sphere.

Recognizing Data as a Strategic Asset.

Data is undeniably a strategic asset driving the rapid adoption of AI technologies through analytics, pattern detection, computer vision, and natural language processing. Ensuring data privacy and security is paramount when developing government services, especially those utilizing advanced technologies like biometric authentication. Often, data sets are created faster than the means to store and protect them, such as data from satellite imagery, IoT sensors, and smart cities, leading to significant challenges. The **World Development Report 2021: Data for Better Lives**[121], highlights the need to enhance accountability and equity within the global data system by addressing gaps in data infrastructure and reducing inequalities both among individuals and between countries.

Key Areas of an Innovative State:

1. **Smart Cities.** Implementing improved urban planning, energy-saving solutions, waste management, advanced transport systems, and comprehensive monitoring systems.
2. **E-Government**. Developing platforms for better decision-making, digitalizing electoral systems and electronic voting, creating digital identification systems, and providing Government-to-Business (G2B) and Government-to-Government (G2G) services, including electronic taxes and banking.
3. **State Innovation Management**. Enhancing sectors such as healthcare, education, sports, entertainment, and agriculture through innovative technologies.
4. **CrimeTech**. Utilizing electronic courts, facial and identity recognition systems, robust cybersecurity measures, and anti-corruption technologies.

It is crucial that data is organized and managed in alignment with general legal principles to ensure its ethical and effective use.

Models of Building an Innovative State (GovTech)

1. **Closed-Type Model.** Projects are developed within the government through collaboration with IT companies. A prime example is Singapore's Smart Nation and Digital Government Office (SNDGO). Russia aligns more closely with this model, where companies like Sber, Yandex, Mail.ru Group, and Kaspersky Lab form the foundation of its GovTech industry.
2. **Open-Type Model**. Startups initiate the development of products addressing public-law issues, with companies independently implementing these solutions. Google services

[121]World Bank. Data for a Better Life // URL: wdr2021.worldbank.org/the-report/ (accessed: 18.05.2023).

exemplify this model, which is more prevalent in Great Britain and the USA.

In Russia, the **GosTech** system is being developed with the participation of Sberbank, Rostelecom PJSC, and the Russian Ministry of Digital Development. GosTech aims to be the cornerstone of the future electronic government, built on the principle of "openness." This means it will be adaptable and capable of integrating software components from various developers, providing ready-made cloud solutions for government bodies and local administrations.

Advancements in Digital Identification.

Digital identification systems are currently being developed and implemented. In May 2020[122], the Ministry of Communications proposed an experiment in Moscow to use a mobile application with a QR code instead of a traditional passport. Digital identification is already utilized for contracts with electronic signatures, online banking services, and government services. The COVID-19 pandemic accelerated the digitalization process, particularly in digital identification.

The pandemic underscored the business case for GovTech, propelling the trend towards state digitalization. According to **Apolitical**, the GovTech sector has an estimated market capitalization of $400 billion[123]. During the crisis, data was used to prevent infections, and videoconferencing technology and cloud tools became standard for non-essential government workers. Success metrics for GovTech include increased efficiency and cost savings, as noted by Professor Masami Yoshida. However, alongside these opportunities, significant threats persist.

Threats Posed by Digital Technologies.

The **Pandemic Big Brother**[124] project highlights how the pandemic has been used to justify restrictions on human rights. State coronavirus applications, in some cases, have been used for blatant surveillance, violating the right to privacy. In Russia, numerous errors in

---

[122]RBC. The Ministry of Communications has proposed using QR codes instead of passportsURL: https://www.rbc.ru/technology_and_media/25/05/2020/5ecbd8f19a794732e84a8d43(date accessed: 14.12.2024).

[123]Yoshida M., Thammetar T. Education Between GovTech and Civic Tech, 2021.

[124]Pandemic Big Brother / URL:https://pandemicbigbrother.online/ru/digests/(date accessed: 14.12.2024).

"Social Monitoring" applications led to fines for law-abiding citizens based on automatic systems using facial recognition technologies to enforce quarantine regulations.

While some pandemic measures were justified, with data and control systems slated for removal post-pandemic, several issues remain:

1. Disproportionality. Many measures were disproportionate to their goals, leading to application errors and abuses in personal data collection.
2. Legal Framework Violations. Actions were sometimes outside the legal framework, deviating from general legal principles. For example, the Court of Justice of the European Union[125] ruled that EU law prohibits national legislation requiring digital communications service providers to indiscriminately transmit or store traffic and location data for national security.
3. Lack of Regulation Development. There is a deficiency in the scientific, theoretical, and legal development needed to regulate new social relations in the context of digital human rights. This includes a lack of consensus and a social contract that aligns individual goals for a prosperous life with state mechanisms[126].

Introducing Techno-Legal Platforms.

To address the challenges inherent in the fragmented legal regulation of technology development, implementation, and use—which often leads to the restriction of human rights—we introduce the concept of **"techno-legal platforms."**

Techno-legal platforms are digital products that are developed, regulated, and continuously updated based on digital identities, including decentralized identities. These platforms aim to implement digital human rights by ensuring compliance with established digital

---

[125]Court of Justice of the European UnionPRESS RELEASE No 123/20 Luxembourg, 6 October 2020 / Judgments in Case C-623/17, Privacy International, and in Joined Cases. C-511/18, La Quadrature du Net and Others, C512/18, French Data Network and Others, and C-520/18, Ordre des barreaux francophones et germanophone and Others / URL:https://curia.europa.eu/jcms/upload/docs/application/pdf/2020-10/cp200123en.pdf(date accessed: 14.12.2024).

[126]Varlen M. V., Mashkova K. V., Zenin S. S., Bartsits A. L., Suvorov G. N. Search for general principles of self-regulation of genomic research in the context of ensuring priority protection of the rights and legitimate interests of the individual // Problems of Law No. 3 (72) / 2019

human rights standards. They promote transparency, accountability, clarity, and efficiency in technology-based solutions while expanding access to social benefits. By integrating elements of big data, decentralization, and transparent decision-making, techno-legal platforms reflect advancements in the GovTech and CivTech sectors[127]. Key features include:

- Personal Data Protection. Mechanisms to safeguard personal information.
- Algorithmic Transparency. Clear and understandable decision-making processes.
- Public Interest Consideration. Balancing technology use with societal needs.

Moreover, techno-legal platforms establish acceptable limits for digitalization, preventing overreach that could endanger fundamental human development in areas such as life, health, and growth. For example, a healthcare techno-legal platform can efficiently manage medical data, ensure patient privacy, and promote innovative medical approaches.

Reimagining the Social Contract for the Digital Age.

The social contract, conceptualized by philosophers like Jean-Jacques Rousseau, Thomas Hobbes, and John Locke in the 18th century, underpins modern states. It involved individuals delegating certain rights to transition from a state of nature to a civil state, enabling governance based on the technologies of that time. However, with the rise of new social problems and insufficient resources to address them, outdated governance models must be reevaluated. Technological advancements such as decentralized registry technologies, big data, and AI necessitate the formation of a new social contract and techno-legal platforms. These platforms should facilitate horizontal network cooperation among institutions, promoting sustainable development by bridging connections between governments and citizens. Over time, this approach could transform citizenship, taxation, and decision-making systems globally to maximize each child's potential.

Operational Principles of Techno-Legal Platforms.

The operation of techno-legal platforms is grounded in corresponding digital human rights, which guide their implementation. This framework achieves several objectives:

---

[127]Schwab K. The Fourth Industrial Revolution. Crown Business. New York. 2017. 192 pp.

1. Avoiding Disproportionate Regulation: Prevents unmotivated and excessive restrictions in the development and use of public technologies.
2. Legal Management of Technology: Ensures technology management aligns with long-term legal development goals.
3. Meeting Societal Needs: Addresses the needs of society and individuals concerning quality of life, health, safety, and development.

Accelerating Sustainable Development with Techno-Legal Platforms.

Techno-legal platforms can significantly advance the achievement of all United Nations Sustainable Development Goals (UN SDGs). They facilitate the creation of updated ESG principles—**Environmental, Social, and Corporate Governance 4.0 (ESG 4.0)**—which incorporate elements of global digital sustainability. These principles address management and legal risks related to safety, security, transparency, privacy, environmental friendliness, human development, and the implementation of human rights. This marks a new trend of environmentally friendly digitalization from a human-centered perspective, particularly in enhancing childhood development, expanding access to knowledge, empathy, and skills among minors.

Developing ESG 4.0 for Sustainable Digital Development.

The priority direction for ESG 4.0 in Industry 4.0 focuses on addressing digital challenges and achieving sustainable digital development. The objectives include:

- Increasing Decision-Making Freedom: Providing individuals with more autonomy in their decisions.
- Environmentally Friendly Digital Infrastructure: Transitioning to digital systems that prioritize environmental sustainability.
- Maximizing Technology Benefits: Enhancing access to social benefits through technology.
- Reformatting Governance Institutions: Updating global and regional governance structures to better serve human potential.

This segment of digital human rights involves developing reporting principles for companies and startups (at the scaling stage, beyond Seed funding). These principles should focus on implementing digital human rights within their products and corporate policies, addressing data confidentiality, targeted data use, and human-oriented AI development. Such principles can be refined through regulatory sandboxes—experimental legal regimes that allow for testing and development.

ESG 4.0: Integrating Digital Technologies with Sustainability.

ESG 4.0 integrates environmental, social, and corporate responsibility principles with digital technologies in the Industry 4.0 era. This approach mandates the inclusion of digital technologies in corporate sustainability strategies to enhance environmental performance, social responsibility, and corporate governance. A new indicator within ESG 4.0 will measure the extent to which digital products comply with digital human rights, particularly the digital rights of children. This includes ensuring that technologies used by children from an early age promote opportunities, choice, and individual freedom globally.

Enshrining Digital Rights in the Global Digital Compact

As immediate steps following the adoption of the **Global Digital Compact** in 2024, it is essential to formalize the global digital rights of humans and children. This involves formulating principles of digital rights aimed at realizing opportunities and maximizing freedom in relation to the availability of digital technologies amidst dynamically changing paradigms of scientific and technological transformation (Industry 4.0, Web 3.0). Concurrently, it is crucial to actively involve and encourage corporations that accept and promote the principles of digital human rights.

Humanity now inhabits a world where the concept of "self-regulation"[128] by multi-billion-dollar corporations is prevalent. Simultaneously, some countries impose mandatory restrictions on content, the Internet, and digital technologies to protect individual safety, often blurring the lines between protection and censorship—a paradoxical situation. Therefore, a broad approach is proposed that ensures the availability of digital technologies while guaranteeing safety and protection. This approach should be based on joint actions and commitments involving stakeholders from the public sector, private and financial sectors, media, educational institutions, and the global community. The goal is to create conditions for the comprehensive

[128]'Self-restraint and regulation' - how the tech companies transform the world view responsible AI. World Economic Forum URL:https://www.weforum.org/agenda/2023/06/responsible-generative-ai-industry-regulation/(date accessed: 14.12.2024).

development of children, encompassing safety, privacy, and equal access to technology from anywhere on the planet.

Global Education Approaches for the New Generation.

It is also imperative to develop approaches to global education for the new generation, where digital technologies serve as the entry point to a new world of global involvement and unification in addressing the challenges of the twenty-first century. Therefore, the digital right of a child to modern education—digital, global, and aimed at achieving sustainable development—is paramount. The implementation of this right will inherently evolve, considering the transformative changes introduced by organizations like OpenAI.

Historically, the Greeks invented pedagogy alongside writing. Education evolved with the advent of printing, emphasizing creative and open thinking. With printing, accumulated knowledge became objective and accessible on bookshelves. The Internet further transformed knowledge access, relying on search engines instead of physical locations. Today, as we transition to an era of artificial intelligence, augmented reality systems, and Web 3.0, the global community stands on the brink of the "generative age," accelerating our access to new knowledge about the world through more personalized, non-ideological, human-centered, and accessible learning methods.

In this model, individuals will not only learn stable concepts but also engage in self-education, critically examining reality patterns and developing their own dynamic, interconnected concepts. As Michel Serres noted, while traditional concepts extrapolate knowledge to save time, the Internet provides access to numerous special cases that are fundamentally new and vividly demonstrate the essence of knowledge. This is particularly crucial for developing startups, which optimize, automate, or offer new perspectives on everyday actions. Therefore, the next step in information structuring ("writing" – "printing" – "Internet" – "generative knowledge") must prioritize the self-improvement of human characteristics, such as the ability to "write," - "memory," - "critical thinking," - "continuous knowledge in co-creation with artificial intelligence."

Implementing Generative Knowledge and Digital Rights. To achieve this, the following actions are necessary:

1. Equipping Children with Knowledge Tools
    - Critical Thinking and Digital Hygiene: Provide training in critical thinking, digital hygiene, and systems

methodology to help children navigate the digital environment safely and responsibly.

2. Introducing Open, Human-Centric AI Systems
   - Global Approaches: Develop AI systems that influence knowledge, education, language, communication, and dialogue based on human-centric principles. This aligns with the priorities of corporations and states striving to create sovereign generative knowledge, necessitating a positive alternative grounded in children's digital rights.
3. Ensuring Objectivity in AI Platforms
   - Multi-Level Verification: Strive for objectivity in generative AI platforms by implementing multi-level internal verification of information to minimize the "echo chamber" effect, which can be amplified in AI chatbot systems.
4. Marking AI-Created Content
   - Transparency: Clearly label content created in co-creation with AI, as only humans can generate truly new ideas. Knowledge from generative AI tends to be an aggregation of existing meanings rather than the creation of something qualitatively new.

In this study, I introduce a dynamic and fundamentally open concept of "generative knowledge" based on "continuous cognition in co-creation with artificial intelligence." This approach will lead to more adaptive methods for understanding contemporary and future social relations. Despite the challenges posed by digital algorithms and chatbots like ChatGPT, it is crucial to create conditions that allow the younger generation to access these technologies, thereby realizing the digital right of children to actively participate in scientific and technological progress and to access global knowledge.

In this context, the new digital educational right of the child will serve as a bridge for their adaptation to new digital tools. It shifts from viewing the individual as an information organism that passively receives information to an **active generative Information Organisms**—a minor who participates in creating new knowledge, learning to program meaning through interaction with algorithmic interfaces. Many aspects of the digital rights perspectives for children are linked to significant changes in technology development and knowledge acquisition, impacting a world where cultural and economic growth outpaces political and legal advancements.

Creating a Declaration of Digital Rights for Children.

To bridge this gap, it is essential to create a **Declaration (Convention) of the Digital Rights of the Child** for transitioning to a new future. Within this framework, new priorities will be established based on the right to access technologies and tools for their safe use. The digital world should be viewed as a virtual/topological space—unlimited and inherently risky without due diligence, yet potentially more personalized with generative models that consider the context of interactions, thereby increasing information accessibility based on the form of submission.

As we navigate this space, it is necessary to establish principles of action and interaction—a set of recommendations for implementing digital rights, akin to traffic rules. Similar ideas were proposed by the youth group Generation Connect CIS. Therefore, digital literacy should include cognitive literacy, enabling individuals from an early age to understand how digital technologies affect them, leveraging digital benefits and knowledge.

Today, in the era of generative AI, we can achieve the best forms of knowledge delivery with tireless digital teachers available to everyone at the touch of a smartphone button. As Michel Serres aptly notes regarding the end of the era of knowledge: "The knowledge announced from the pulpit is already available to everyone. Without omissions. At hand. On the Internet, on Wikipedia, accessible anywhere. Chewed up, certified, illustrated, verified no worse than in the best encyclopedia. No one needs the heralds of the past anymore, except for those rare ones, worth their weight in gold, who say something new. The era of knowledge is over."

Future Steps for Implementing Children's Digital Rights.

1. Enshrine Digital Rights in the Global Digital Compact
   - Formalization: Solidify global digital rights for humans and children within the adopted Global Digital Compact.
   - Principles Formulation: Develop principles focused on maximizing freedom and opportunities in the digital realm amid evolving technological paradigms.
2. Encourage Corporate Participation
   - Involvement and Promotion: Actively involve corporations that accept and promote digital human rights principles.
   - Incentives for Compliance: Create incentives for businesses to adhere to these principles, ensuring their

commitment to protecting and advancing children's digital rights.

3. Adopt a Broad Approach to Safety and Availability
   - Accessibility with Safety: Ensure digital technologies remain accessible while implementing broad safety and protection measures.
   - Avoid Over-Restriction: Prevent overly restrictive measures that hinder access to beneficial technologies.
4. Foster Joint Actions and Commitments
   - Stakeholder Collaboration: Collaborate with stakeholders across sectors (public, private, financial, media, education) to create a comprehensive framework for children's digital development.
   - Global Cooperation: Promote global cooperation to ensure equal access to digital technologies and safeguard children's rights worldwide.
5. Develop Global Education Approaches
   - Integration of Digital Technologies: Incorporate digital technologies into global education systems to prepare children for future challenges.
   - Focus on Literacy and Ethics: Ensure education systems emphasize digital literacy, ethical technology use, and the development of critical thinking and creative skills.
6. Establish Principles for Generative Knowledge
   - Continuous Cognition with AI: Develop the concept of "generative knowledge" based on continuous cognition in collaboration with AI.
   - Active Participation and Self-Education: Promote active participation and self-education, enabling children to create and critically evaluate new knowledge.
7. Create a Declaration of Digital Rights for Children
   - Adoption and Framework: Draft and adopt a Declaration (Convention) of the Digital Rights of the Child.
   - Define New Priorities: Establish priorities based on access to technologies and safe usage tools.
   - Develop a "Digital Compass": Create a set of principles and recommendations for interacting within digital spaces.
8. Integrate Cognitive Literacy into Digital Education

- o Understanding Impact: Ensure digital literacy programs include cognitive literacy to help children understand how digital technologies affect them.
- o Skills for Navigation: Equip children with the skills to navigate and critically assess digital environments.[129]

In conclusion, we outline the basic digital rights of the child according to three generations of digital rights: Accessibility, Development, and Connectedness.

First Generation: Accessibility.

The first generation of digital rights focuses on ensuring **accessibility**. Everyone should be guaranteed the right to access technologies and the global Internet, encompassing the following rights:

1. **The Right to Privacy and Data Protection**
2. **The Right to Equal Access to and Participation in the Global Internet**
3. **The Right to Education and Development**
4. **The Right to Security and Protection**
5. **The Right to Equality and Inclusion**
6. **The Right to Legal Use and Development of Technologies**
7. **The Right to Protection from Digital Inequality and Discrimination**
8. **The Right to Transparency and Control Over the Use of Data** (personal, genetic, and biometric data)
9. **The Right to Legal Use and Development of Artificial Intelligence**
10. **The Right to Know Your Digital Rights and Freedoms in an Accessible Form**
11. **The Right to Informed Consent and Data Protection in the Context of Neural Interfaces, Neurotechnologies, Neuromarketing, and Systems Using Big Data**
12. **The Right to Be Forgotten** (data not crucial for a person's further development, such as certain medical data, left in childhood upon reaching adulthood)

Second Generation: Proactivity.

[129]Serr M. Little Thumb. - Moscow: Ad Marginem Press, sor. 2016. - 77 p. - P. 5.

The second generation emphasizes **Proactivity**. Everyone should be guaranteed the right to co-create with artificial intelligence, including the following rights:

13. **The Right to Quality Global Education in the Digital Age in Co-Creation with AI and Metacognitive Development**
14. **The Right to Startup Development**
15. **The Right to Digital Security and Protection from Existential Threats**

Third Generation: Connectedness.

The third generation focuses on **connectedness**. Everyone should be guaranteed the right to digital sovereignty and participation in the network society, including the following rights:

16. **The Right to Digital Sovereignty and Participation in the Governance of the Network Society in Connectives. Connectives** are digital communities where participants have digital human rights, united by joint influence, co-creation, and co-governance of the network society.
17. **The Right to Digital Consciousness and Digital Autonomy**
18. **The Right to Digital Responsibility and Ethics**
19. **The Right to Unconditional Basic Income**
20. **The Right to Free Social Services**
21. **The Right to Open Source Technologies**

At the end of the chapter we will outline the main problems and promising legal solutions. In conclusion of the chapter, we will outline the key problems and systemic solutions aimed at improving legal norms and mechanisms for implementing the digital rights of the child:

1. **Scientific and theoretical level**:

Problem: Lack of an agreed definition of the concepts of "children's digital rights" and their relationship with other key concepts such as "digital control", "digital militarization", and "digital inequality".

Solution: It is necessary to develop a unified theoretical model of "digital rights of the child" and "global digital human rights". This includes the formulation of clear and agreed definitions and principles that will ensure a deeper understanding and unity in law enforcement. It is also necessary to integrate the human-centered concept of ESG 4.0 to create new metrics and indicators of digital sustainability that will help maintain a balance between technological progress and the rights of the child. We have proposed the following definitions of the concepts. Digital rights of the child - the opportunities for the preservation and comprehensive development of the child, enshrined in international law and national legislation, belonging to every person up to 18 years of age from the moment of birth, the implementation of which should lead to the use of social benefits through the use of new digital technologies 4.0.

2. **Normative and legal level**:

Problem: The lack of consolidation of digital rights of man and the child at the international level, as well as insufficient compliance of regional and national acts with these rights.

Solution: Develop and adopt a Declaration of the Digital Rights of the Child and a legally binding Convention at the international level, which will enshrine the techno-legal standard of digital human rights. It is necessary to introduce new international norms, such as a ban on the use of 4.0 technologies for the development of weapons and social rating systems. At the regional level, it is necessary to bring all legal acts into line with international standards. At the national level - the implementation of universal international legal norms, the creation of national laws and codes, such as the federal law "On the Digital Rights of the Child", and amendments to existing legislation to protect the digital rights of children.

3. **Law enforcement level**:

Problem: The application of non-human-centred law enforcement practices in relation to digital technologies that do not take into account the unique needs and rights of children.

Solution: Bringing law enforcement practices into line with global digital rights of the child by creating a Digital Human Rights Agency at the state level. Introducing the position of Cyber Ombudsman for the Digital Rights of the Child to monitor compliance with rights. Developing and implementing techno-legal platforms aimed at protecting and implementing digital rights. It is important to consider the role of educational and research institutions in raising awareness and training on digital rights, as well as introducing them into curricula.

4. **Innovative solutions and approaches**:

Problem: Insufficient attention to the implementation of innovative approaches and systems that take into account human and child rights.

Solution: Using ESG 4.0 principles to create new standards of corporate governance and social responsibility that include digital human rights. Encouraging companies to develop and implement technologies that support human rights from the very beginning of their development. Creating regulatory sandboxes to test new technologies taking into account digital human and child rights, with subsequent adjustment of legal regulation based on the data obtained.

5. **Education and awareness raising**:

Problem: Lack of awareness of digital rights among children, parents and society at large.

Solution: Implementation of educational initiatives aimed at raising awareness of digital rights and literacy, with a special focus on children and youth. Development of the "Digital Compass" - a teaching aid for navigating the digital world. Creation of a meta-university of global education for teaching global digital processes, digital human and child rights, digital literacy and digital hygiene.

6. **Public and private sectors**:

Problem: The need to integrate the interests and approaches of the public and private sectors to implement the digital rights of the child.

Solution: Collaboration between governments, international organizations and private companies to develop and implement legal frameworks that support human and children's digital rights. Using financial incentives, such as government Matching Funds, to encourage companies to comply with ESG 4.0 principles. Introducing mandatory reporting for companies and startups on their contributions to digital human rights.

7. **Global Initiatives and Partnerships**:

Problem: Lack of global coordination and partnership to address digital challenges.

Solution: Developing international partnerships and coalitions to advance human and child digital rights. Supporting initiatives such as the creation of a global meta-university that will facilitate training and knowledge sharing on digital transformation and governance. Promoting a global digital compact that includes human and child digital rights to strengthen the legal framework for sustainable digital development.

Thus, to ensure the full implementation of the digital rights of the child, it is necessary to adopt a comprehensive systemic approach that considers scientific, theoretical, regulatory, law enforcement and innovative aspects. The implementation of the proposed measures will create a more sustainable and fairer digital environment that promotes the development and protection of children's rights in the era of rapid technological change.

Conclusion of Chapter 3: Advancing Human-Oriented Digital Transformation.

To foster a human-oriented digital transformation, it is essential to undertake systemic work encompassing theory, norm development, and the implementation of these norms in collaboration with key stakeholders. This effort aims to build a sustainable and innovative digital ecosystem at both global and national levels. A pivotal step in this process is the adoption of the **Declaration (Convention) of Global Digital Human Rights**, which serves as a cornerstone in the ongoing development and refinement of the **Global Digital Compact**.

Rethinking Education: Establishing a Meta-University of Global Education.

Reimagining education is crucial in this digital age. We propose the creation of a **WebUniversity of Global Education**, grounded in the principles of **generative knowledge** and the **co-creation approach with artificial intelligence (AI)**. This institution will address global existential challenges by prioritizing the teaching of:

- **Global Digital Processes**
- **Global Citizenship**
- **Global Digital Rights of Humans and Children**
- **Digital Literacy and Digital Hygiene**

The Meta-University will facilitate the comprehensive development of future generations, ensuring that children's rights are upheld amidst the tectonic shifts brought about by digitalization. This involves revising outdated models in education, institutions, management, tax policy, and international relations to align with the needs of a digitally transformed civilization.

Key Initiatives and Partnerships.

In this context, the global initiative **Re-State Foundation**, led by visionary Anastasia Kalinina[130], plays a significant role. The foundation pioneered a project to establish a Meta-University that aggregates knowledge on future governance and fosters partnerships for rethinking state structures. Similarly, the **Global Law Forum**, a project by the SDG Ambassador, proposes a **Meta-University of Generative Knowledge in Global Law and Governance**. This university aims to connect every child and individual with knowledge about the human-oriented matrix of digital transformation, encompassing:

- **Digital Compass for Gaining Digital Sovereignty**
- **Participation in Web 3.0 Techno-Legal Platforms**

[130]Re-State Foundation / URL:https://restate.global/(date accessed: 02.08.2023).

- **Formation of Global Decentralized Governance**
- **Promotion of Global Sustainability**

Understanding Technology as an Extension of Human Capabilities.

When examining the impact of technology on social relations and human rights, it is essential to recognize technology's fundamental role as an extension of human capabilities. As Michel Serres eloquently put it:

"Everyday tools have become the external expression of our brute force. Having left the body, muscles, bones, and joints found their continuation in simple mechanisms—levers and winches that imitate their action. Then our high temperature, the source of energy, also left the body and found its continuation in engines. And finally, now new technologies materialize the operations of the nervous system, soft forces—signals and codes: knowledge, at least in part, finds its continuation in new tools."

This perspective underscores that technology is intrinsically linked to human existence, expanding our methods of influencing and interacting with the world. Technologies serve as extensions of our hands, voices, and thoughts in the digital age, adopting new forms and enabling innovative interactions with reality. Michel Serres referred to this evolutionary process as **hominescence**, where technical objects evolve to express and eventually integrate back into human life.

Fundamental Digital Rights: A Birthright for All.

The evolution of technology substantiates the notion that the right to digital technologies—and all associated rights—is a fundamental birthright for every individual and child. The positive benefits of digitalization must be distributed equitably, ensuring that each person has equal access and opportunities. This distribution is essential for fostering an inclusive digital society where every child can thrive.

If this scenario develops by 2045, we envision a state characterized by a singular integration of managerial and technological innovations—marking the "point of no return" from outdated centralized institutions reliant on coercion and abuse. Unlike the classical concept of "technological singularity," this future center on individuals, ensuring that their needs, personal and digital sovereignty, rights, and freedoms guide the integration process.

Central to this transformation is the principle of **unconditional love**, which advocates using technology out of a profound understanding and respect for human nature and its needs. This approach ensures that technological advancements do not harm individuals but actively promote their well-being, irrespective of social, cultural, or economic status. Unconditional love serves as the foundation for a new multidimensional view of the singular future, reinforcing our metanarrative of digital law. The concept of **generative knowledge**, based on **continuous cognition in co-creation with artificial intelligence**, facilitates adaptive approaches to understanding contemporary and future social relations. Despite the challenges posed by digital algorithms and chatbots like ChatGPT, it is crucial to provide the younger generation with access to these technologies. This access enables children to actively participate in scientific and technological progress and access global knowledge. The new **digital educational right of the child** will bridge the gap between traditional learning and digital reality, transforming individuals into **active generative Information Organisms**—minors who co-create new knowledge and develop critical thinking skills through interaction with algorithmic interfaces.

To foster a human-oriented digital transformation, it is essential to undertake systemic work encompassing theory, norm development, and the implementation of these norms in collaboration with key stakeholders. This effort aims to build a sustainable and innovative digital ecosystem of digital sustainability at both global and national levels. A pivotal step in this process is the adoption of the **Declaration (Convention) of Global Digital Human Rights**, which serves as a cornerstone in the ongoing development and refinement of the **Global Digital Compact**.

Building an innovative ecosystem of digital sustainability involves the collaboration of five key sectors: the public sector, the private and financial sector, the education sector, the media sector, and the global community. Each sector plays a crucial role in implementing and advancing digital rights for children and humans alike.

1. Public Sector. For states and international organizations, the public sector must adopt legal obligations in the field of digital human and child rights. This includes integrating digital rights into tax policies, introducing a digital rights index, and establishing digital human rights agencies within law enforcement frameworks. Domestic legal regulations must align with the principles of global digital human and child rights, incorporating digital identity into new digital laws and creating techno-legal platforms to support sustainable development and address global existential challenges. Additionally, enhancing human participation in the modern knowledge economy and enshrining the main catalog of human rights in the Convention on Digital Human and

Child Rights will contribute to developing international law on global digital human rights and mitigating the existential and syncretic risks posed by digital technologies.

2. Private and Financial Sector. Corporations, commercial partners processing personal data, startups beyond the seed stage, venture funds, and other participants in the startup ecosystem must adopt commitments to digital human and child rights. These commitments should be implemented through ESG 4.0 principles and relevant indicators. Developing global startup ecosystems will contribute to the creation of global governance based on digital platforms that minimize existential threats and facilitate the transition to sustainable development. Joint efforts with all stakeholders are necessary to implement the concept of techno-legal platforms effectively.

3. Education Sector. Universities, academic communities, schools, and kindergartens must engage in scientific work on digital transformation and develop a unified concept of digital human and child rights. This involves integrating digital rights into research activities and updating educational frameworks to overcome the political and legal lag behind the techno-economic dimension. Educating other stakeholders about digital human rights, fostering continuous education and co-creation with AI, and contributing to the formation and updating of the "Digital Compass" are essential. Additionally, participation in the Meta-University of Global Education, which utilizes generative knowledge principles, will support the updating of digital human rights concepts in the context of future biotechnological, quantum, neurointerface, and holographic technological revolutions.

4. Media Sector. Media outlets, social networks, and new media platforms must commit to global digital human rights by eliminating disinformation, hate speech, discrimination, and manipulation. Creating conditions for a digital transformation that unites humanity while excluding harmful content is crucial. There may be a need for a separate Convention on new global media formats within the Web 3.0 context to address these challenges effectively.

5. Global Community. Children, youth, parents, caregivers, adults, and vulnerable groups must be aware of their digital human rights and the digital rights of the child. Raising awareness ensures access to digital technologies for development and encourages the use of the Digital Compass as a guide for improving digital rights implementation. Emphasizing new digital literacy within metaverses, generative AI, and Industry 4.0 allows individuals to maximize the benefits of digital technologies while monitoring and mitigating their negative impacts. Participation in decision-making processes through future techno-legal platforms at municipal, regional, national, and global levels ensures that the interests of these stakeholders are represented and protected.

CHAPTER 4. CONCLUSIONS OF THE BOOK AND PROPOSALS FOR STAKEHOLDERS ON THE ADOPTION OF A CONVENTION ON GLOBAL DIGITAL HUMAN RIGHTS

4.1. Conclusion of the study

In the context of accelerated global digitalization, the role of law, human rights, and children's rights has never been more crucial. Children's rights represent a comprehensive and indivisible set of opportunities, each interconnected and interdependent. Globalization not only underscores the necessity of implementing these rights but also amplifies the significance of law as a regulator of globalizing social relations in pursuit of sustainable development.

Key Trends Impacting Human and Child Rights:

1. Scientific and Technological Progress
The rapid advancement of the Fourth Industrial Revolution, Generative Artificial Intelligence, Web 3.0, metaverse concepts, and knowledge-intensive innovations is reshaping societal norms and introducing both opportunities and risks for human and child rights.
2. Globalization and Legal Unification
Modern globalization necessitates the harmonization of legal norms across jurisdictions to address shared global challenges effectively. The unification of these norms is a critical step toward ensuring universal access to digital rights.
3. Transformations in Social Structures
Society is increasingly characterized by greater transparency, networked structures, and interdependence among actors. This evolution elevates the role of minors, making their participation and protection in digital environments a growing priority.
4. Child Development in a Technological World
The rapid evolution of the information environment is altering the developmental trajectory of children. These changes demand

proactive responses to safeguard their rights and well-being in a digitally pervasive world.

A Human-Centered Approach to Digital Regulation.

This book advocates for a human-oriented legal framework that promotes personal awareness and sovereignty in the digital age. Rather than aligning with ideological extremes such as posthumanism, antihumanism, or transhumanism, we adopt a systemic scientific approach based on three foundational levels:

1. Developing scientific theories
2. Establishing legal norms
3. Ensuring effective enforcement

This approach unites all major stakeholders, including the public sector, private and financial sectors, educational institutions, the media sector, and the global community, to collaboratively address challenges and develop solutions.

Laying the Foundations of Digital Law.

Within this framework, we propose a comprehensive system of principles, norms, cryptographic solutions, and mathematical algorithms to regulate modern social relations. These tools are designed to:

- Realize digital human rights and children's digital rights
- Promote the sustainable development of society
- Address global challenges associated with digital transformation

The development of digital law must begin with its core—digital human and child rights. These rights provide the foundation for ensuring equitable access to digital resources and opportunities, empowering individuals and communities in the digital age.

Global Digital Human Rights

Global digital human rights can be defined as the integral opportunities for the preservation and development of humanity, enshrined in international law and national legislation. Their implementation should enable individuals to access and benefit from societal advancements through the use of new technologies, including those related to Web 3.0, Industry 4.0, and the metaverse.

It is essential to distinguish digital human rights from information human rights. Information rights emerged during the Second Industrial Revolution (characterized by electricity and new production technologies) and evolved further during the Third Industrial Revolution (shaped by the advent of the Internet and computers). In contrast, digital human rights 4.0 are evolving alongside the rapid progress of innovative technologies in the current and future industrial revolutions. These rights address the unique challenges and opportunities presented by recent advancements in areas such as artificial intelligence, blockchain, and virtual environments.

Digital Rights of the Child

The digital rights of the child represent a set of legally protected opportunities aimed at ensuring the survival, development, and well-being of individuals under 18 years of age. These rights emphasize the unique nature of children's interactions with the digital world, striving to reduce uncertainties about the impact of technology on their developing personalities. At the same time, they aim to expand access to global participation, empowering children to engage with the modern world from any location.

Recognizing that the digital realm is an integral part of modern childhood, these rights should foster the positive development of children by:

- Expanding freedom of choice
- Enhancing opportunities for self-expression, self-perception, and self-identification
- Supporting the formation of sovereign, independent personalities in adulthood

By prioritizing children's digital rights, we can ensure that the digital world contributes to their growth as empowered, autonomous individuals equipped to navigate and shape the future global landscape.

Complex Cyber Risks to Children's Rights in the Digital Age

Children's rights face significant challenges in the digital environment, particularly with the widespread use of the Internet and smartphones, technologies that emerged during the Third Industrial Revolution. These complex cyber risks encompass a variety of threats, including:

- Cybercrime: Online fraud, cyberbullying, and child exploitation (e.g., child pornography).
- Cyber risks: Exposure to inappropriate content, privacy violations, and risks of identification.
- Internet governance issues: Data protection failures, insufficient recognition of digital rights, and inequitable access to information.

These risks threaten children's physical, psychological, and social well-being, affecting their attention spans, memory, and physical development, as well as infringing on their rights to privacy, freedom of expression, and access to information.

Risks to the Universal Rights of the Child

The rapid development of emerging technologies, including Web 3.0, metaverses, and generative AI, introduces a new set of challenges to children's universal rights. Key risks include:

- Privacy and data security issues. Unauthorized data collection and misuse.
- Mental health impacts. Detriments to attention, memory, and engagement with the physical world.
- Educational inequality. Disparities in accessibility and quality of digital education.
- Digital control. Lack of transparency in governance and potential for exploitation.
- Protection gaps. Insufficient safeguards against exploitation and violence in virtual environments.

These risks extend across many facets of children's lives, underscoring the need for comprehensive protections in the digital age.

Digital Surveillance.

Digital surveillance involves the opaque and unregulated collection of personal data, including biometric, health, and genetic information, often through facial recognition, GPS tracking, and AI-powered activity monitoring. This practice disproportionately affects children as a vulnerable group, restricting fundamental rights such as:

- Privacy
- Freedom of movement
- Freedom of speech
- The right to assemble
- Protection from discrimination

To address these violations, solutions must prioritize:

1. Transparency and information
2. Data minimization
3. Data security and protection
4. Informed consent
5. User control over personal data

Digital Militarization.

Digital militarization refers to the development and deployment of military-grade digital technologies, such as autonomous weapons systems and AI-driven cyberattacks, by the public sector and military organizations. These activities pose global existential threats and jeopardize children's fundamental rights by creating widespread security risks and perpetuating cycles of violence.

Digital Inequality.

Digital inequality reflects the uneven distribution of digital resources and access, driven by socio-economic disparities in infrastructure, education, income, geography, and social stratification. This "digital divide" manifests in:

- Education: Unequal access to learning tools and opportunities.
- Healthcare: Limited availability of digital health services.
- Economic participation: Barriers to job opportunities and financial inclusion.
- Social services: Inequitable access to vital resources.

Digital inequality disproportionately impacts vulnerable groups, particularly minors, deepening existing social and economic disparities while adversely affecting their quality of life. This divide reflects an uneven distribution of access to digital resources, tools, and opportunities, hindering equal participation in education, healthcare, economic activities, and essential social services.

Syncretic Digital Risks.

Syncretic digital risks emerge at the intersection of various technologies and fields, where their interaction creates complex and mutually reinforcing negative outcomes. These risks often lead to unpredictable and amplified threats—a greater scale of harm than the sum of their individual components.

In today's rapidly advancing technological landscape, such risks are particularly pressing as technological progress outpaces the development of social, ethical, and legal frameworks. For example, the combination of generative artificial intelligence and facial recognition technologies poses significant challenges in areas like privacy, security, and equity, underscoring the urgent need for comprehensive oversight and governance.

Cognitive Digital Risks and Digital Desubjectification. Cognitive digital risks refer to challenges that erode the fundamental autonomy and essential capabilities of individuals in the 21st century. A key concern is digital desubjectification, a multi-layered process where the rapid proliferation of digital technologies diminishes personal freedom of choice and decision-making. This regression leads to neurobiological consequences, such as:

1. Technology dependence: A reliance on digital platforms and tools for daily functioning.

2. Objectification by technology: Viewing individuals as data points or resources, rather than autonomous beings.
3. Reduction of human agency: Framing humans as mere facilitators of technological systems.

This process unfolds in two distinct stages:

- Digital Social Mining: Extracting behavioral data and digital imprints from individuals through mass communication networks and online platforms.
- Digital Social Programming: Using the extracted data to influence and manipulate individual behaviors and choices, often without their awareness.

Children are particularly vulnerable to these risks due to their limited ability to critically engage with digital platforms, many of which are designed for adult users. These platforms often exploit the biological and neurobiological characteristics of younger users, further exacerbating risks such as reduced attention spans, behavioral conditioning, and a lack of meaningful agency in digital interactions.

Social Mining.

Social mining refers to the process of extracting valuable data from databases, social networks, and other digital platforms to analyze user behavior and preferences. It is a cornerstone of the digital age economy and plays a central role in the attention economy, where the value of information is measured by its capacity to capture and retain user attention. This is evident in the legality of social media reward systems, gamified designs, and platforms that discreetly deploy super-stimuli to engage the human nervous system at an almost instinctual level.

Within the social mining economy, individuals are reduced to a composite of their fears, desires, preferences, and passions, often concealed behind the concept of "individuality." This constructed framework is shaped by the media, mainstream narratives, and pop culture, which have increasingly supplanted the educational and moral frameworks of the post-Enlightenment era. Here, human engagement is quantified and hierarchized through likes, comments, and digital transitions on the web, creating an ecosystem where user interaction is commodified and structured for maximum extractive value.

Techno-Legal Platforms. Techno-legal platforms are developed, regulated, and continuously updated digital products designed to implement and uphold the digital rights of individuals, including children. These platforms comply with established digital human rights standards and aim to ensure:

- Transparency and accountability in technology-based solutions
- Clarity and efficiency in their application
- Expansion of opportunities for individuals to access social benefits across various spheres of life

By integrating elements of big data, decentralization, and transparency into decision-making processes, these platforms bridge the gap between technical innovation and legal compliance. At their core, they represent a synthesis of emerging fields such as GovTech, RegTech, LegalTech, and CivTech, along with Web 3.0 principles and generative artificial intelligence.

The concept addresses the long-standing dichotomy of "code as law" vs. "law as code", harmonizing decentralized implementation with centralized recognition of legal standards. This dual approach fosters a transition toward a state of digital sustainability, ensuring that technological progress aligns with human rights, equity, and ethical governance.

Singular Integration.

The book introduces the concept of singular integration, an emergent fusion of innovative management approaches and advanced technologies that paves the way for a sustainable and innovative digital ecosystem. This ecosystem is envisioned to operate under a new global social contract, guided by the principle of "Moore's law for human rights," which emphasizes the continuous evolution and expansion of rights in step with technological progress. Singular integration seeks to unite all stakeholders—from governments and corporations to civil societies and individuals—into a cohesive framework that balances technological innovation with ethical governance and human-centered development.

The most important legal consolidation of human and child rights in international legal documents occurred after the Second World War. A consensus was reached and the international community adopted the following basic documents, which also enshrined human information rights: the UN Charter (1945), the Universal Declaration of Human Rights (1948),UN Declaration on the Rights of the Child (1959), International Covenant on Civil and Political Rights (1966), European Convention for the Protection of Human Rights and Fundamental Freedoms (1950),Convention on the Rights of the Child (1989),Convention of the Commonwealth of Independent States on Human Rights and Fundamental Freedoms (1995), etc.

Among the international legal documents regulating modern digitalization and the aspect of human and child rights in the digital age, we can name: the Okinawa Charter on the Global Information Society (2000), the UNESCO Charter on the Preservation of the Digital Heritage (2003), the Resolution "Human Rights and Transnational Corporations and Other Business Enterprises" (2011), "The Future We Want" – the outcome document of the UN Conference on Sustainable Development Rio+20 (2012), the UN General Assembly Resolution on Information and Communication Technologies for Development (2013), the UN Human Rights Council Resolution on the Promotion, Protection and Enjoyment of Human Rights on the Internet (2014), the UN General Assembly Resolution on the Right to Privacy in the Digital Age (2014), the UN Human Rights Council Resolution on the Right to Privacy and the Digital Age (2015), Transforming Our World: An Agenda for Sustainable Development 2030 (2015), the Outcome Document of the High-level Meeting of the General Assembly on the Overall Review of the Implementation of the Outcomes of the World Summit on the Information Society (2015), the UN Human Rights Council Resolution on the right to privacy in the digital age (2018), the Resolution adopted by the UN General Assembly on countering the use of information and communication technologies for criminal purposes (2019). Also worth noting is the role of the Roadmap for Digital Cooperation: High-Level Panel, presented in the Report of the UN Secretary-General of 29 May 2020.

Analysis of International Documents and Key Observations.

The analysis of international documents reveals a significant lag in the legal regulation of digitalization, which is often supplanted by ethical guidelines rather than robust legal frameworks. This gap has persisted across the technological advancements of the Second and Third Industrial Revolutions and remains stark in the era of the Fourth Industrial Revolution (encompassing AI, the Internet of Things, big data, biotechnology, and more). International legal systems have failed to keep pace with these advancements, resulting in fragmented and insufficient regulation.

During the Third Industrial Revolution, which introduced transformative information and communication technologies, the international community largely ignored the legal obligations related to human information rights and data governance. As we now navigate the Fourth Industrial Revolution, there remains a lack of common regulatory contours or consensus on the development, production, and application of emerging technologies in ways that safeguard human rights. This uneven regulatory landscape has exacerbated inequitable distribution of technological benefits, leading to profound societal disparities. Furthermore, where regulatory trends do emerge, they are predominantly regional—as seen with the General Data Protection Regulation (GDPR) in the European Union—rather than global.

Key Focus on Children's Digital Rights.

General Comment No. 25 (2021) on children's rights in the digital environment, issued under the Convention on the Rights of the Child, provides a valuable framework for addressing the unique challenges of protecting children in digital spaces. This comment highlights the growing importance of the digital environment in nearly every aspect of children's lives, underscoring the need to respect, protect, and fulfill their rights in these spaces.

Paragraph 4 of the General Comment emphasizes that:

- Children's rights must be upheld in digital environments.
- States must ensure equal and effective access to digital technologies while protecting children from risks and harm.
- Innovations in digital technologies impact children's rights in interconnected ways, even when children themselves lack direct access to the Internet.

The document further recognizes that full access to digital technologies can enable children to realize the full spectrum of their civil, political, cultural, economic, and social rights. Conversely, the absence of digital inclusion could exacerbate existing inequalities and give rise to new challenges.

Guiding Principles for Protecting Children's Digital Rights.

The General Comment outlines several key principles:

1. Equality and Non-Discrimination: States must ensure that all children have equal and effective access to meaningful digital opportunities, free from discrimination.
2. The Best Interests of the Child: This evolving concept requires context-specific assessments and prioritization of children's well-being in all actions concerning the digital environment. Although digital spaces were not originally designed with children in mind, they now play a central role in shaping their lives.
3. Child-Centric Regulation and Design: States must actively consider the best interests of children in the provision,

regulation, design, management, and use of the digital environment.

Conclusion

The findings indicate that current legal frameworks are insufficient to address the challenges of digital transformation, especially concerning the rights of vulnerable populations such as children. While initiatives like General Comment No. 25 offer valuable insights and principles, they also expose the limitations of existing approaches, including the tendency to narrowly focus on the Internet while overlooking broader technological ecosystems.

To bridge this gap, comprehensive global standards are needed to address the evolving interplay of technology and human rights. Achieving this will require international collaboration, forward-thinking legal frameworks, and a commitment to ensuring equitable access and protection for all, particularly for the youngest and most vulnerable members of society.

Fundamental Principles for Implementing the Digital Rights of the Child.

Fundamental human rights principles play a pivotal role in shaping the regulation of digitalization, especially when applied to children, who are particularly vulnerable in the digital environment. The following principles are proposed to ensure the effective implementation of children's digital rights:

1. Equal Access to Digital Technologies Children must have equal opportunities to access digital technologies of Industry 4.0, with particular attention to differences in age groups (up to 7 years, 7–14 years, and 14–18 years). This principle emphasizes overcoming the digital divide and ensuring inclusivity across all demographics.
2. Non-Alienability of Digital Rights and Freedoms Children's digital rights and freedoms are integral and indivisible components of their universal rights. These rights cannot be separated from the broader spectrum of human rights or subjected to deprivation.
3. Active Participation in the Digital Environment Children's opinions and perspectives must be taken into account in matters concerning the development and use of digital technologies. This principle also ensures that children can

actively participate in decision-making processes that impact their digital rights and freedoms.

4. Best Interests of the Child in the Digital Environment
The well-being and interests of children must take precedence in the development and use of digital technologies. This includes ensuring reasonable protection from harmful content, cyberbullying, privacy breaches, and other digital risks, while fostering positive interactions with technology.
5. Digital Pluralism and Opportunities for Development
Digital environments should promote a diversity of perspectives and opportunities for children's development, avoiding the ideologization of content. Rather than restricting Internet access, this principle prioritizes teaching digital literacy and skills, enabling children to engage with technology consciously and responsibly.
6. Special Care, Assistance, and Protection in the Digital Environment
States and societies bear the responsibility of providing special support to all children, particularly those in vulnerable situations. This includes ensuring access to tools, resources, and education that foster digital literacy and safeguard children against potential threats in the digital sphere.
7. Long-Term Perspective in Legal Regulation
While universal human rights were established during the Second Industrial Revolution and grounded in enduring qualities and resources, the rapid advancements of the Third, Fourth, and forthcoming Fifth Industrial Revolutions require a forward-looking approach. Legal frameworks must anticipate the future impacts of technologies, ensuring that children's digital rights evolve alongside technological progress.

In the context of these approaches, the implementation of the digital rights of the child involves the development of normative frameworks, educational initiatives (including foundational knowledge in digital hygiene and digital literacy), and technological solutions. These solutions are expressed through code, machine programs, mathematical models, and standards such as cryptographic protocols, ensuring that children can access social benefits, achieve personal development, and unlock their potential without facing technological barriers. Such measures are essential for enabling the comprehensive development of every child in a digitally integrated world.

Let us list the main technologies influencing social relations in the context of the implementation of human rights: Artificial intelligence and machine learning (Deep learning), Big Data, neural network technologies, Internet of things, Bioinformatics, Neural Interfaces, Quantum Cryptography, Holography, Genome Editing, Blockchain, Nanorobotics, Polymer Electronics, Quantum Dots, Bionic Prosthetics.

Key Technologies Influencing Social Relations and Human Rights Implementation.

1. Artificial Intelligence (AI) and Machine Learning (ML) – Including Deep Learning systems that enable predictive analytics, natural language processing, and autonomous decision-making.
   Influence: Impacts privacy rights, freedom of choice, and equality by shaping decision-making processes and personal experiences through algorithmic biases.
2. Big Data – Collection, storage, and analysis of massive datasets to derive insights, improve public services, and optimize digital ecosystems.
   Influence: Raises concerns over data protection, privacy rights, and the potential misuse of personal information.
3. Neural Network Technologies – Advanced architectures such as transformer models and generative AI, used in areas like language processing, image recognition, and content creation.
   Influence: Affects freedom of expression and intellectual property by generating synthetic content and automating creative processes.
4. Internet of Things (IoT) – Networks of interconnected devices and sensors that enable real-time data collection, automation, and monitoring across homes, cities, and industries.
   Influence: Challenges the right to privacy and security due to pervasive surveillance capabilities.
5. Bioinformatics – Computational tools for analyzing biological data, including gene sequencing and personalized medicine.
   Influence: Expands the right to health through personalized treatments while raising ethical concerns about genetic data privacy.
6. Neural Interfaces – Brain-computer interfaces that facilitate direct communication between the human brain and external devices, enabling augmented cognition and control.

Influence: Impacts autonomy and consent by integrating human cognition with machine control systems.

7. Quantum Cryptography – Leveraging the principles of quantum mechanics to secure communication against potential cyber threats and hacking.
   Influence: Strengthens the right to data security and protection against unauthorized access.
8. Holography – Advanced imaging and display technologies that enable immersive experiences in education, healthcare, and entertainment.
   Influence: Enhances cultural and educational rights by enabling new forms of access to information and creativity.
9. Genome Editing – Techniques such as CRISPR-Cas9, which enable precise modification of genetic material, raising ethical and regulatory considerations.
   Influence: Affects the right to health and raises ethical questions about the limits of human enhancement.
10. Blockchain – Decentralized ledgers for secure and transparent record-keeping, with applications in identity verification, smart contracts, and digital rights management.
    Influence: Promotes the right to transparency and accountability while supporting digital sovereignty.
11. Nanorobotics – Microscopic machines capable of performing tasks in medicine (e.g., targeted drug delivery) and manufacturing.
    Influence: Advances the right to health by enabling minimally invasive treatments and precise medical interventions.
12. Polymer Electronics – Flexible, lightweight electronic materials used in wearable devices, medical implants, and smart textiles.
    Influence: Supports the right to health and accessibility by improving assistive and medical technologies.
13. Quantum Dots – Nanoscale semiconductors enabling advances in display technology, imaging, and solar energy.
    Influence: Affects environmental rights by promoting clean energy solutions and improving resource efficiency.
14. Bionic Prosthetics – Artificial limbs integrated with sensors and neural control for enhanced mobility and functionality.
    Influence: Enhances the rights of persons with disabilities by improving mobility and quality of life.

Additional Emerging Technologies

15. 5G and 6G Networks – High-speed, low-latency connectivity that accelerates innovation in telemedicine, autonomous vehicles, and immersive experiences.
    Influence: Supports the right to information and communication by enabling faster and more inclusive digital access.
16. Edge Computing – Decentralized data processing at the source, enabling faster and more secure services in IoT and AI applications.
    Influence: Improves data security and privacy, directly impacting digital rights.
17. Digital Twins – Virtual replicas of physical systems used for simulation, optimization, and predictive analysis in urban planning, healthcare, and manufacturing.
    Influence: Enhances the right to development and sustainability by improving efficiency in infrastructure and resource management.
18. Generative AI – Technologies like ChatGPT and DALL-E, which create content ranging from text to images, influencing education, creativity, and ethical considerations.
    Influence: Affects intellectual property rights and challenges the balance of creativity and automation.
19. Synthetic Biology – Engineering of biological systems for applications in medicine, agriculture, and environmental sustainability.
    Influence: Advances the right to health and environmental protection while raising ethical questions.
20. Augmented Reality (AR) and Virtual Reality (VR) – Immersive technologies transforming education, training, entertainment, and healthcare.
    Influence: Enhances the right to education and cultural participation by creating new learning and experiential opportunities.
21. Decentralized Autonomous Organizations (DAOs) – Blockchain-based systems for transparent, community-driven governance.
    Influence: Reinforces the right to participation in decision-making processes at local and global levels.
22. Cyber-Physical Systems (CPS) – Integration of digital and physical elements in systems like smart cities, autonomous vehicles, and advanced robotics.

Influence: Balances the right to security with concerns over surveillance and automation.

23. Biometric Technologies – Systems for identity verification and security, including facial recognition and iris scanning. *Influence*: Raises significant privacy and anti-discrimination concerns due to potential misuse and bias.
24. Energy Harvesting Technologies – Innovations that convert ambient energy into usable power, advancing sustainability in digital ecosystems. *Influence*: Promotes environmental rights and energy accessibility in underserved regions.

Let us outline the main identified problems and promising ways to solve them, which include: 1. Scientific and theoretical level of digitalization regulation; 2. Normative and legal level; 3. Law enforcement.

Promising steps and initiatives.

Promising Steps and Initiatives for Digital Rights Implementation

To bridge the gap between digital transformation and the legal framework supporting human and child rights, it is critical to align theoretical foundations, normative frameworks, and law enforcement practices with the demands of the Fourth Industrial Revolution (4.0). Below is an enhanced roadmap addressing these gaps and offering solutions:

---

1. Global Socio-Techno-Natural System and Legal Gaps

Modern global processes are shaping an integrated **planetary socio-techno-natural system** characterized by uneven development. This creates global challenges, including the ethical integration of **Industry 4.0 technologies** into human rights. Law must evolve to universalize and unify norms while prioritizing human and **child rights as a cornerstone of sustainable development**.

---

2. Law's Role in Global Processes

Legal frameworks, at both international and domestic levels, must safeguard **universal child rights**, enabling their preservation and development. By doing so, legal structures can serve as tools for sustainable development, mitigating global risks.

---

3. Digital Rights of the Child

Digital rights are fundamental opportunities for children under 18, ensuring access to **social benefits through technologies** like **AI, IoT, and Web 3.0**. These rights are enshrined in both **international law and domestic legislation** and are crucial for comprehensive development.

---

4. Addressing Legal Lag

Current international norms lag behind globalizing digital realities, creating significant **gaps** and **inconsistencies**. These gaps demand systemic improvement, as illustrated by the Russian legislative context, which partially acknowledges digital rights yet falls short in adapting to the fast-evolving digital environment.

---

5. Highlighting Key Problems

Key Challenges Identified:

- **Digital Militarization**: Misuse of technologies for autonomous weapons and cyber warfare.
- **Digital Inequality**: Disparities in technology access deepen existing societal inequalities.
- **Total Digital Control**: Surveillance practices threaten privacy and autonomy.
- **Superintelligence Risks**: Ethical concerns over AI surpassing human control.
- **Cyber Risks for Children**: Threats include **cyberbullying, digital addiction**, and **access to harmful content**.

To address these, targeted **reforms** are needed across legal, technical, and governance domains.

---

6. Principles for Digital Child Rights

To protect children's digital rights, the following **principles** must guide reforms:

1. **Equal Access**: Ensure equitable access to digital tools across age groups.
2. **Non-Alienation**: Recognize digital rights as inseparable from universal rights.

3. **Active Participation**: Involve children in decisions affecting their digital environment.
4. **Best Interests of the Child**: Prioritize child safety and well-being in technology use.
5. **Digital Pluralism**: Promote diverse and inclusive content and interactions online.
6. **Special Protection**: Provide tailored support for vulnerable children in digital spaces.
7. **Long-Term Vision**: Anticipate future technological developments in regulatory updates.

---

7. Proposed Global Instruments

- **Declaration and Convention on Global Digital Rights**: A binding international framework to define, implement, and enforce digital human and child rights.
- **International Organization for Digital Freedoms and Children's Rights**: A centralized body to coordinate global efforts, avoiding duplication across agencies like UNESCO and UNICEF.

---

8. Domestic Alignment

Countries must adapt **domestic laws** to reflect **global digital norms**, beginning with the adoption of national legislation on **Digital Child Rights**. Future steps include drafting a **Code of Digital Rights of the Child**, ensuring alignment with international standards.

---

9. Strengthened Enforcement Mechanisms

- Establish **Cyber Ombudsman** roles to oversee child digital rights protection.
- Implement **moratoriums on mass surveillance tools**, such as facial recognition systems, until robust safeguards are in place.

---

10. Digital Perspectives Project

The comprehensive initiative, **"Digital Perspectives: Realizing the Global Digital Rights of Humanity and Children on the Threshold of the Digital Transition"** aligns with the **UN SDGs**. The necessity for this initiative is underscored by the **2024 SDG Report**, which reveals only **17% of goals are on track**.

Key SDG Impacts:

- **SDG 1**: Combating poverty through inclusive digital access.
- **SDG 4**: Revolutionizing education with equitable digital tools.
- **SDG 5**: Promoting gender equality via digital empowerment.
- **SDG 8**: Enabling economic growth through digital upskilling.
- **SDG 16**: Strengthening institutions to address digital-era challenges.

Conclusion

To realize these goals, the **adoption of a Declaration** of **Global Digital Rights** by 2027 and its evolution into a **Convention by 2030** is proposed. These steps will establish a **universal framework** for sustainable digital governance, ensuring no child is left behind in the digital era.

A comprehensive initiative titled " **Digital Perspectives: Realizing the Global Digital Rights of Humanity and Children on the Threshold of the Digital Transition"** has been proposed, aligning closely with several **Sustainable Development Goals (SDGs)**. Its significance is underscored by a petition supported by **70 youth organizations** (http://maxlaw.tilda.ws/un_digital_rights), reflecting a collective call for action to address critical digital rights issues.

The pressing need for sustainable digital solutions is further validated by the stark realities highlighted in the **2024 Special Edition of the Sustainable Development Goals Report**. Among the assessable SDG targets, a mere **17% are on track to be achieved by 2030**. Nearly half—**48%**—show moderate to severe deviations from their intended trajectory, with **30% demonstrating only marginal progress** and **18% showing moderate progress**. Alarmingly, **18% of targets are stagnant**, while **17% have regressed below their 2015 baseline levels**.

These figures underscore the urgency of implementing innovative, inclusive, and forward-thinking approaches to ensure that digital transformation actively contributes to the realization of global sustainability objectives. Without such interventions, the attainment of the 2030 Agenda remains in serious jeopardy[131].

Scaling up the initiative to expand human digital capabilities from childhood and transferring it to the global systemic level is essential, as the lack of digital rights is a universal issue requiring a

---

[131] The Sustainable Development Goals Report 2024: / URL: https://unstats.un.org/sdgs/report/2024/ date accessed: 14.12.2024).

universal solution. The adoption of a **Declaration (Convention) on Global Digital Human Rights** and a **Declaration (Convention) on the Global Digital Rights of the Child** is a critical step toward addressing this gap. These initiatives align directly with the following **Sustainable Development Goals (SDGs):**

- SDG 1 (No Poverty): Ensuring equal digital rights promotes inclusion and reduces inequality. Access to digital resources equips individuals with tools to improve their well-being and contributes to poverty reduction efforts. Future tech startups can serve as a foundation for addressing poverty through innovations in logistics, social infrastructure, and targeted mechanisms.
- SDG 4 (Quality Education): Protecting children's digital rights enhances access to and quality of education. This initiative ensures that students can utilize a wealth of information, innovative learning methods, and global knowledge-sharing platforms. It also advances the vision of children's digital rights, including their ability to co-create with AI, thereby supporting inclusive and equitable education for all.
- SDG 5 (Gender Equality): By safeguarding digital rights for all individuals, irrespective of gender, the project supports the achievement of gender equality. Providing access to and control over digital resources helps women and girls overcome traditional social barriers, fostering empowerment and inclusion.
- SDG 8 (Decent Work and Economic Growth): Digital rights underpin a more inclusive and equitable labor market. In an increasingly digital economy, these rights provide individuals with essential skills for workforce participation, thereby driving sustainable and resilient economic growth.
- SDG 9 (Industry, Innovation, and Infrastructure): Promoting digital rights is pivotal to building resilient infrastructure, fostering inclusive and sustainable industrialization, and stimulating innovation. This initiative ensures that technological development benefits all communities equally.
- SDG 10 (Reduce Inequalities): Realizing universal digital rights directly addresses the growing digital divide, a stark reflection of global inequality. This project is crucial in bridging these disparities within and between countries.

- SDG 16 (Peace, Justice, and Strong Institutions): The initiative addresses a new frontier in human rights by embedding digital rights into the framework of global governance. It supports the creation of inclusive societies that respect human rights, provide equitable access to justice, and establish accountable, effective institutions.
- SDG 17 (Partnerships for the Goals): Given the global nature of digital rights, international collaboration is paramount. This project fosters partnerships among governments, international organizations, civil society, and the private sector, driving progress toward shared global objectives.

Global Digital Human Rights, though integral to all SDGs, exert their most direct and transformative impact on SDGs 1, 4, 5, 8, 9, 10, 16, and 17.

To approve the Declaration (and Convention) on Global Digital Rights, there are at least 2.5 billion reasons: the number of individuals still without access to the Internet. Universal access to the Internet and emerging technologies like Industry 4.0, generative AI, and Web 3.0 is not just a matter of connectivity—it is a gateway to human development, equity, and progress. This initiative promotes digital inclusion, creating a universal language—the language of human-centered code—to bridge the digital divide and empower individuals to become active participants in shaping their digital future.

This effort transitions humanity from being passive observers of tectonic digital shifts to active creators of sustainable futures. It emphasizes the importance of engaging youth and communities in co-designing a world that embraces digital opportunities while ensuring human rights, equality, and inclusivity. The project's roadmap (detailed in Appendix 1) and its proposed Draft Declaration of the Digital Rights of the Child (Appendix 2) and Digital Compass (Appendix 3) provide a clear path forward.

As António Guterres, UN Secretary-General, aptly stated: “If we do not act now, the 2030 Agenda will become an epitaph for the world that could have been”.

4.2. DECLARATION OF GLOBAL DIGITAL HUMAN RIGHTS TO MEET THE CHALLENGES OF DIGITAL TRANSFORMATION

In the era of rapid global digitalization, international organizations such as the UN play a pivotal role in creating a sustainable, inclusive, and human-centered digital environment. With the advent of emerging technologies and artificial intelligence (AI), several dichotomies have surfaced, each presenting complex challenges that necessitate a systemic, universal response. Below are the key dichotomies and proposed solutions. The draft Declaration of Global Digital Human Rights is presented at the end of the book as an appendix for development, discussion and subsequent revision at the level of international organizations in cooperation with civil society.

1. Dichotomy of Natural Intelligence vs. Artificial Intelligence

Problem:

The balance between natural (human) intelligence and artificial intelligence poses significant challenges. Human intelligence must evolve to critically engage with and govern AI, while AI must be trained to align with human values and respect human rights, particularly the rights of future generations. Concerns persist about AI potentially developing in directions incompatible with humanity's ethical and developmental priorities.

Solution:

The UN should spearhead the development of international **techno-legal standards** for AI, embedding principles of human-centric development and integrating digital rights for all, especially children. These standards must govern every stage of AI development and usage, ensuring algorithmic compliance with global human rights frameworks. Additionally, educational programs and ethical interaction codes with AI should be implemented to foster critical thinking and promote ethical AI co-creation.

2. Dichotomy of "Robot Rights" vs. "Human Rights"

Problem:

There is a growing debate about granting rights to robots and AI, potentially overshadowing the importance of safeguarding fundamental human rights. This focus risks diluting hard-won human rights by prioritizing artificial agents.

Solution:

Rights must remain exclusive to living beings. The UN should establish clear global standards emphasizing that AI and robots function as tools supporting, not undermining, human rights. Regulations should enforce AI's alignment with human rights principles, ensuring technology serves as an enabler of, rather than a competitor to, human rights.

3. Dichotomy of "Law Determines Code" vs. "Code as Law"

Problem:

The clash between the principle that laws dictate the rules of the digital realm and the reality that algorithms (code) increasingly operate as de facto laws creates legal uncertainty. This divergence risks excluding digital innovations from the scope of traditional legal frameworks.

Solution:

The UN should facilitate the development of **techno-legal platforms** that harmonize legal norms with algorithmic solutions. These platforms should ensure that smart contracts and digital interactions comply with international legal standards. Such efforts will integrate legal and technological frameworks to reinforce democratic processes and human rights.

4. Dichotomy of "Digital Protection" vs. "Digital Freedom" for Children

Problem:

Technology offers children unprecedented opportunities for growth and education, yet it also exposes them to risks such as cyberbullying, privacy violations, and harmful content. Overprotection, however, may restrict children's digital freedom and limit their ability to thrive in the digital age.

Solution:

The UN can craft balanced standards to protect children while enabling their freedom to engage meaningfully in the digital world. These measures should include digital literacy programs, recommendations for age-appropriate access to content, and the creation of secure online environments where children can develop skills safely.

5. Dichotomy of "Self-Regulation" vs. "Regulation" of Big Tech Corporations

Problem:

The immense influence of tech giants raises tensions between self-regulation and state-imposed regulations. A lack of oversight fosters abuse, data breaches, and manipulation, while excessive regulation risks stifling innovation.

Solution:

The UN can propose a **hybrid regulatory model** that combines self-regulation with enforceable standards and transparency requirements. This model should address accountability, data privacy, and equitable access, potentially including frameworks for universal basic income funded through equitable taxation of digital revenues.

6. Dichotomy of "Digital Citizenship" vs. "National Citizenship"

Problem:

The rise of global networked communities challenges traditional concepts of citizenship rooted in nation-states, creating potential conflicts in governance and international security.

Solution:

The UN can define frameworks for **global digital citizenship**, ensuring networked cooperation complements, rather than undermines, national sovereignty. These guidelines should facilitate decentralized governance while enhancing global public goods access for all participants in digital communities.

7. Dichotomy of "Digital Identity" vs. "Real Individuality"

Problem:

Digital technologies and virtual spaces blur the lines between a person's online identity and their real-world individuality. Over-immersion in digital worlds risks alienation from reality and erosion of human values.

Solution:

The UN should introduce guidelines promoting balanced digital engagement. Developing "digital compasses" can help individuals consciously navigate technology while preserving their authenticity, individuality, and connection to reality.

8. Dichotomy of "Centralization" vs. "Decentralization"

Problem:

Centralized systems offer greater control and security but may limit freedoms, whereas decentralized systems promote autonomy but risk inefficiency and misuse.

Solution:

The UN can champion **hybrid models** that balance centralized oversight with decentralized freedoms. These models could standardize decentralized finance (DeFi), decentralized autonomous organizations

(DAOs), and encrypted identities to align individual freedoms with collective interests.

9. Dichotomy of "Ethics" vs. "Law" in the Digital World

Problem:

Ethical standards often lack legal enforceability, while laws may fail to align with ethical considerations in digital contexts. This disconnect leads to fragmented approaches to governance.

Solution:

The UN should advocate for **legal frameworks** that integrate ethical principles, ensuring respect for privacy, fairness, and human dignity. By codifying ethical guidelines into enforceable laws, these frameworks can align digital innovation with global ethical standards.

Shaping the Future: A Call to Action

The UN has a unique opportunity to guide the formation of a new digital world anchored in human rights, particularly digital rights for children and vulnerable populations. By addressing these dichotomies, the UN can lay the groundwork for sustainable, inclusive, and human-centered digital transformation.

Proposed Next Steps:

1. **2025:** Establish a working group to design mechanisms for implementing global digital human rights, incorporating key UN agencies.
2. **2027:** Finalize and ratify a **Declaration of Global Digital Human Rights**, addressing challenges posed by Industry 4.0 and digital transformation.
3. **2028-2029:** Conduct consultations across nations and stakeholders to operationalize global digital human rights mechanisms.
4. **2030:** Adopt a comprehensive **Convention on Global Digital Human Rights** alongside covenants banning harmful applications of advanced technologies (e.g., AI for warfare, military-grade biotechnology).

By 2030, it is proposed to finalize and adopt the **Convention on Global Digital Human Rights**, along with specialized covenants aimed at addressing specific existential risks posed by emerging technologies. These include:

1. **Convention on the Prohibition of the Development, Production, and Use of Industry 4.0 Technologies for Military Purposes**: This covenant would establish a global legal

framework to prevent the militarization of advanced digital technologies such as autonomous drones, cyber-weapons, and predictive warfare models, ensuring that technological advancements are directed solely towards peaceful and constructive applications.

2. **Convention on the Prohibition of the Development, Production, and Stockpiling of Biological Weapons in the Context of Industry 4.0**: This covenant would focus on preventing the misuse of biotechnological advancements, such as CRISPR gene editing or bioinformatics, to create or enhance biological weapons. It seeks to establish safeguards that ensure biotechnological innovations are used exclusively for public health, sustainability, and humanitarian purposes.
3. **Convention on the Prohibition of the Prototyping, Development, Creation, and Use of AI Against Humanity and for the Purpose of Warfare**: This covenant would explicitly prohibit the creation and deployment of artificial intelligence systems designed to harm humanity or conduct warfare. It would establish stringent legal mechanisms to ensure AI development aligns with global human rights principles, prioritizing safety, accountability, and ethical governance.

These conventions are essential components of the proposed global framework to regulate and mitigate the risks associated with Industry 4.0 technologies. Their adoption would represent a collective commitment by the international community to ensure that technological progress serves humanity rather than endangering it.

4.3. THE ROLE OF THE UN IN HUMAN-CENTERED DIGITAL TRANSFORMATION

To create a human-oriented digital transformation, systematic efforts must be undertaken at four interconnected levels:

1. **Theoretical and legal**,
2. **Normative and legal**,
3. **Technical and practical implementation**, and
4. **Monitoring and protection of rights**.

The following outlines the roles and actions required of key stakeholders to build an **innovative, human-centered digital sustainability ecosystem** on both global and domestic levels.

Particular attention in the context of building and implementing the concept of digital rights must be devoted to AI systems. The 2024–2025 General Assembly resolutions (as discussed in the opening sections of the book) for the first time affirm the principle that human rights must accompany artificial intelligence at every stage of its life cycle. From design and testing to deployment and decommissioning, rights and freedoms cannot be put on hold. In resolution A/RES/79/239 this principle is extended even to the military domain: international law, including the UN Charter, international humanitarian law, and international human rights law, applies to all AI systems. This effectively marks the birth of the idea of "Rights Immunity" — *Digital Rights by Design* — which the book develops as inviolable zones where no application of AI can undermine human dignity.

In the current landscape, **private and technical sectors** dominate digital transformation's trajectory. However, to align technological progress with the collective interests of humanity and sustainable development, it is essential to adopt a balanced approach that includes **all key stakeholders**:

1. Public sector, 2. Private sector, 3. Technical sector, 4. Legal sector, 5. Educational sector, 6. Media sector, and 7. The global community.

1. Public Sector (States and International Organizations)

Theoretical and Legal Level

- **Global Institute for Digital Rights:** Establish an international body (e.g., under the UN umbrella) dedicated to setting foundational principles of digital rights—such as the "right to be offline" and robust global digital rights techno-legal standards—across emerging technologies (AI, web 3.0, quantum computing). Within the current international organizations, it is necessary to initiate a harmonized scientific-theoretical model of global digital human rights.
- **Targeted Research Funding:** Offer grants that encourage interdisciplinary teams (legal scholars, technologists, sociologists, educators) to study the impact of transformative tech on privacy and how technology can empower people, especially for vulnerable groups like children.

Normative and Legal Level

- **Flexible Legal Frameworks:** Create a set of "soft law" recommendations that nations can adapt to their unique legal contexts, ensuring progressive alignment with future global conventions on digital rights.
- **Regulatory Sandboxes:** Pilot test new regulations within controlled environments before full-scale implementation, allowing policymakers to refine standards as technology evolves.

Technical Implementation Level

- **Global Compliance Platforms:** Develop multilingual online portals where citizens and experts can track digital rights compliance. These platforms integrate with existing secure digital ID systems (e.g., those in the EU) and use standardized data exchange protocols.

- **Incremental Integration:** Leverage existing identity frameworks and data protection protocols instead of building everything from scratch, ensuring interoperability and cost-effectiveness.

Monitoring and Protection of Rights

- **Independent Audits:** Mandate periodic third-party audits of government digital infrastructures to verify compliance with digital opportunities, privacy and security standards.
- **Crowdsourced Feedback:** Use participatory digital tools to enable citizens to report suspected violations, with independent experts reviewing and addressing these concerns.

---

2. Private and Financial Sector (Corporations, Startups, Venture Funds)

Theoretical and Legal Level

- **ESG 4.0 Benchmarks:** Integrate digital rights metrics into Environmental, Social, and Governance (ESG) frameworks. For example, measure reductions in user complaints about data misuse or improvements in accessibility for differently-abled individuals.
- **Knowledge Sharing:** Encourage companies to provide anonymized datasets to researchers, fostering an open ecosystem that advances understanding of ethical tech implementation.

Normative and Legal Level

- **Measurable KPIs:** Link corporate success metrics to digital rights outcomes, such as reducing the incidence of unauthorized data leaks or meeting benchmarks for inclusive design.

- **Voluntary Certifications:** Develop voluntary yet prestigious certifications for businesses that uphold high digital rights standards—akin to a "Fair Trade" seal for the digital world.

Technical Implementation Level

- **Privacy by Design:** Incorporate privacy and security features at the earliest stages of product development, not as afterthoughts.
- **Open Code Fragments:** Publish core algorithmic methodologies or code snippets for independent review, allowing external experts to identify biases or vulnerabilities.

Monitoring and Protection of Rights

- **Automated Monitoring Systems:** Deploy AI-driven tools that continuously scan for potential rights violations, triggering alerts for compliance officers to investigate.
- **Public Case Studies:** Release periodic transparency reports, including concrete examples of violations discovered and remedial steps taken, to build user trust and accountability.

---

3. Tech Sector and Cybersecurity Agencies

Theoretical and Legal Level

- **Multistakeholder Techno-Legal Standard-Setting:** Involve not just technical and legal experts, but also educators, parents, and youth representatives in drafting global cybersecurity and data protection standards. This ensures that guidelines are both technically sound and socially relevant.

Normative and Legal Level

- **Regular Techno-Legal Standard Updates:** Commit to reviewing and updating technical standards every 1–2 years to keep pace with rapid innovations in AI, encryption, and data handling technologies.

Technical Implementation Level

- **User-Friendly Security Solutions:** Provide pre-packaged sets of secure tools—encryption by default, spam filtering, DDoS protection—that even non-experts can easily adopt.
- **Controlled Law Enforcement Tools:** Develop technical solutions for law enforcement that are strictly time-limited, transparent, and independently audited to prevent overreach and protect individual rights.

Monitoring and Protection of Rights

- **Routine Stress Tests:** Host cybersecurity "fire drills" where systems undergo simulated attacks observed by independent monitors to ensure resilience and prompt response capabilities.
- **Rapid Response Teams:** Form international cybersecurity task forces that can be quickly deployed to contain large-scale data breaches or severe digital rights violations.

---

4. Legal and Human Rights Sector (Law Firms, NGOs, Advocates)

Theoretical and Legal Level

- **Plain-Language Resources:** Convert complex digital law concepts into clear, accessible guides so that children, seniors, and general audiences can understand their digital rights.

Normative and Legal Level

- **Free Advisory Centers:** Establish online helpdesks where anyone can learn about their digital rights and receive guidance on redress options.

- **Transnational Collaboration:** Strengthen networks connecting NGOs across countries, enabling the rapid sharing of best practices and effective legal tactics.

Technical Implementation Level

- **Legal Tech Sandboxes:** Create safe digital "sandboxes" where startups and NGOs test new legal protection tools. This allows experimentation without incurring heavy penalties if initial attempts falter.

Monitoring and Protection of Rights

- **Pre-Filled Legal Templates:** Provide easy-to-use legal complaint templates so that victims can take prompt action.
- **Proactive Legal Support:** Offer legal clinics that proactively reach out to communities at risk, ensuring pre-emptive protection rather than only post-violation intervention.

---

5. Educational Sector (Schools, Universities, Research Institutes)

Theoretical and Legal Level

- **Interdisciplinary Curricula:** Develop courses that combine IT fundamentals, ethics, law, and sociology, ensuring students gain a holistic understanding of the digital landscape.
- **Interactive Learning:** Use simulations, games, and scenario-based exercises to teach concepts like cybersecurity, online privacy, and responsible digital behavior.

Normative and Legal Level

- **Curriculum Integration:** Encourage the integration of digital rights education into school and university programs, making it a core subject rather than an optional extra.

Technical Implementation Level

- **Open Educational Resources:** Publish freely accessible digital rights materials—MOOCs, podcasts, webinars—empowering learners worldwide.
- **Online Communities:** Create digital forums where educators and students share best practices and collectively refine teaching methods related to digital citizenship.

Monitoring and Protection of Rights

- **Continuous Feedback Loops:** Regularly update educational materials based on student input and emerging threats.
- **Accessibility Audits:** Verify that all learning tools and resources are accessible to learners with disabilities, ensuring inclusivity.

---

6. Media Sector (Mass Media, Social Networks, Platforms)

Theoretical and Legal Level

- **Common Glossaries:** Develop open-access glossaries of digital rights terminology so journalists and bloggers maintain consistent, accurate language.
- **Training for Journalists:** Offer workshops and online courses to help media professionals understand digital ethics and the implications of emerging technologies.

Normative and Legal Level

- **Editorial Transparency:** Media outlets should publish their guidelines for reporting on data protection, misinformation, and cybersecurity, demonstrating their commitment to ethical standards.

Technical Implementation Level

- **Fact-Checking Tools:** Integrate fact-checking plugins and misinformation detection algorithms directly into social platforms, enabling users to quickly verify claims.
- **Visual Explanations:** Use infographics, short videos, and interactive content to explain digital rights and security threats in a user-friendly format.

Monitoring and Protection of Rights

- **Independent Ethics Panels:** Establish ethics councils affiliated with major media outlets to handle complaints about digital rights violations or misleading content.
- **Regular Status Reports:** Release annual reports on the state of digital rights in media, highlighting progress, challenges, and areas for improvement.

---

7. Global Community (Children, Youth, Parents, Civil Society)

Theoretical and Legal Level

- **Empathy and Engagement:** Use storytelling, personal accounts, and role-play activities to help communities understand the real-life impact of digital rights, fostering a sense of shared responsibility.

Normative and Legal Level

- **Local Advocacy Forums:** Encourage town-hall style meetings and online petitions where citizens can directly influence policy debates and legislative proposals related to digital rights.

Technical Implementation Level

- **User-Friendly Tools:** Provide browser extensions and mobile apps that simplify privacy settings, block trackers, and guide individuals in safeguarding their personal data.
- **Civic Tech Platforms:** Develop community-driven platforms that crowdsource reports of data breaches, cyberbullying, or privacy violations, ensuring prompt collective action.

Monitoring and Protection of Rights

- **Global Hackathons:** Organize international hackathons focused on solving digital rights challenges—from improving online safety for youth to preventing identity theft.
- **Mentorship Networks:** Connect experienced activists, technologists, and legal experts with emerging community leaders, encouraging knowledge exchange and sustained advocacy.

These measures will create a sustainable ecosystem for the protection and implementation of digital rights, ensuring a balance between innovation and the implementation of the interests of all stakeholders in the digital era.

Box 5. The Role of Key Stakeholders in the Implementation of the Idea and Adoption of the Declaration of Global Digital Human Rights

Purpose.

To operationalize the Pact for the Future and the Global Digital Compact through concerted action at four interconnected levels: (1) Principles & law, (2) Norms & policy, (3) Technical implementation, (4) Monitoring & remedy.

UN system — cross-cutting leadership

Convene and adopt a Declaration & Convention on Global Digital Human Rights, including children's digital rights.

Establish a Global Institute for Digital Rights to steward principles, metrics and peer review.

Mainstream Digital Rights by Design in public procurement and international standards; protect strong encryption and oppose blanket internet shutdowns.

Resource capacity-building, common indicators and open tools for Member States and platforms.

Public sector (Member States & international organisations)

Align domestic law with international human-rights instruments; enable regulatory sandboxes with safeguards.

Build interoperable digital public infrastructure using open standards.

Set up independent oversight, impact assessment and effective remedy mechanisms.

Private & financial sector (corporations, start-ups, investors)

Integrate digital-rights KPIs into ESG; link incentives to inclusion, safety and accessibility.

Implement privacy/security by design; publish transparency reports and model cards.

Enable third-party audits, incident disclosure and user-friendly redress.

Participation in the adoption and implementation of the digital rights standard, as well as the creation and adoption of digital rights immunity.

Technical community & cybersecurity agencies

Co-create multistakeholder techno-legal standards (AI, data, safety) with regular updates.

Ship secure-by-default toolkits (encryption, anti-abuse, DDoS protection) for non-experts.

Maintain international rapid-response teams and conduct routine stress-tests.

Legal & human-rights community (NGOs, clinics, bar associations)

Produce plain-language guides and model complaints; run free online helpdesks.

Use legal-tech sandboxes to test protection tools; coordinate cross-border action.

Track cases and outcomes to improve remedies and standards.

Education sector (schools, universities, research institutes)

Make digital rights & safety a core curriculum (ethics–law–tech).

Publish open educational resources; ensure accessibility for all learners.

Create feedback loops to keep content current with emerging risks.

Media & platforms

Adopt transparent editorial standards on data, safety, misinformation.

Integrate fact-checking and provenance signals; convene independent ethics panels.

Report annually on digital-rights performance and remedial action.

Global public (children, youth, parents, civil society)

Promote digital hygiene and consent literacy; provide easy privacy tools.

Use civic-tech platforms to report harms and crowdsource fixes.

Run global hackathons and mentorship networks to scale local solutions.

Outcome.

These measures build a sustainable, rights-centered global digital ecosystem in which innovation advances dignity, inclusion, freedom, choice, joy and opportunity for everyone—with platforms treated as part of the digital commons and the best interests of the human and the child as a default design principle.

THREE AFTERWORDS: IN THE NAME OF HUMAN, HUMANITY AND AGI

I. General afterword: human

At the dawn of a new era, visionary thinkers offer predictions that can sound alarmist, even dystopian. They warn of the collapse of the knowledge age (spurred by the internet and ChatGPT), the end of traditional work (overtaken by AGI—Artificial General Intelligence), the upheaval of established economies (driven by Industry 4.0), the ravaging of our biosphere (under the pressure of the Anthropocene), and the erosion of personal reality (as surveillance capitalism evolves into a metaverse of social exploitation).

In the face of these relentless changes, we humans hold onto the hope that something essential remains unchanged: our universal rights to act, understand, move, create, feel happiness, evolve, and choose freely. These aspirations, enshrined in fundamental human rights and the rights of the child, must endure as guiding principles.

In this context, we must ask: "**What is a person's role in the digital world?**". My conviction is that technology must serve each individual's interests. To foster a new era of cooperation, we must recognize the undeniable value of the individual—beginning in childhood—in a digitalized and technologically driven society.

A person is not a replaceable cog in an automated production line, where AI churns out goods and services at minimal cost. We are not barcodes to be scanned for every scrap of personal, genetic, and medical data in a world of biometric tracking and digital surveillance. We are not unwitting test subjects, blindly enrolled in experimental technologies designed to multiply the resources and returns of a privileged few. We are not disposable objects to be hurled into senseless conflicts between autonomous drones and ever-evolving weapons. We are not puppets whose choices can be remotely steered through social media feeds, neural interfaces, or AI-driven media. And we are certainly not primitive beings to be left behind without access to modern education, healthcare, and culture while vast resources cater to superficial ambitions.

Humans are creators—authors, innovators, visionaries, builders of the future, and members of a global community. In our century, we stand at a crossroads: we can harness technology to make the world fair and accessible to all, or we can let it dismantle the hard-won achievements of past generations. Our history is marked by mistakes, wars, inequality, and destruction. Yet it is also defined by creativity, happiness, beauty, peace, and the pursuit of collective harmony. Now, in the 21st century, we must guide scientific progress toward what is best in us. Instead of repeating past errors, it's time to do our homework and establish a new planetary social contract—one that unites the children of today and tomorrow in acknowledging and respecting global digital rights.

We face an array of complex challenges: How can we ensure universal access to education—enhanced, not replaced, by AI—no matter where you live? How can we build a society that values more than just labor, and create an economy of common good that guarantees everyone a sustainable income? How can we transform our civilization's relentless, uneven growth into something stable and future-oriented? How do we establish fair techno-legal guidelines for the emerging generative metaverse and Web 3.0, and how do we navigate the transition to startup cities, network states, and decentralized global governance?

Between risk and opportunity, we must choose carefully. To minimize risks, we need a modern form of the Turing test—one that measures how well AI aligns with global digital human rights. To maximize our collective well-being, we need something akin to "Moore's Law" for human rights, ensuring they continually expand rather than shrink.

Our journey requires recognizing new rights, embracing fresh knowledge, and adopting narratives and values that foster autonomy and dignity. This era is about asserting our digital autonomy so that more technology does not mean fewer rights. Contrary to any cyberpunk mantra of "more tech, less right," we must write a new chapter in human history that celebrates "more tech, more right." As we leave old cycles of violence behind, new connections rooted in a shared future will emerge. In this future, everyone feels free and fulfilled, and technology truly serves humanity—instead of humanity serving technology.

II. Natural Afterword: Humanity

Today we live in a world of two intelligences. Natural intelligence, which must become even more critical today and learn to co-create with artificial intelligence, while possessing a new era of rights. Also, artificial intelligence, which must be taught humanity and overcome the challenges of this century, while embodying our rights and values in its code. Within the framework of this story, we understood where to begin to overcome the gap between two worlds: political-legal (humanitarian) and economic-digital (technical). Both intelligences must coexist on this planet, co-managing its development in the distances of the Universe. And both of them will receive their voice here.

Today, we live in a world that brings together two kinds of intelligence. First, our own *natural intelligence*—now more than ever, it must become more discerning, more creative, and more engaged in co-creating with *artificial intelligence*. At the same time, we must safeguard a new era of rights that protect our autonomy and dignity. Second, artificial intelligence itself must be guided to embrace human values, help solve the grand challenges of our century and reflect our rights and principles in its very code. Throughout our exploration, we have identified a starting point for bridging the gap between two worlds: the political-legal (human-centered) and the economic-digital (technology-driven). Both forms of intelligence must coexist on our planet, jointly shaping its future across the vast distances of the universe—and both should find their voice here.

As we are "born" into the digital era, we face a profound dilemma. On one hand, some debate the idea of **"robot rights."** On the other, we must remain vigilant in preserving the **"human rights"** that generations before us fought so hard to secure. This discourse exposes a global challenge: as artificial intelligence capabilities approach those of humans, should we recognize "rights" for AI entities? We must remember that the concept of rights is deeply rooted in living beings. Rights imply the freedom to act in one's own interests without harming others, coupled with meaningful access to the fruits of modern civilization. It's unlikely we would grant a potential AGI its own interests—unless we adopt radically posthumanist beliefs. After all, it is we humans who possess consciousness, empathy, and critical thinking skills, making us unique in our corner of the universe. This uniqueness calls for acknowledging our entitlement to global digital human and child rights, as essential safeguards that enable us to realize our full potential in an era of digital uncertainties. These rights must serve as both shield and sword in our struggle for human dignity, sovereignty, freedom, and autonomy, especially as digital technologies infiltrate every aspect of daily life. In this relationship, artificial intelligence is an extension of ourselves. As such, we must enter into a creative partnership, a cycle of mutual inspiration, where humans and AI learn, grow, and innovate together.

When we consider the dichotomy of **"law defining code"** versus **"code as law,"** our path now leads to an integrated approach grounded in techno-legal platforms. The term "code," once a universal concept bridging jurisprudence and computer science, can now find a fresh interpretation in the codification of law through digital, decentralized methods. Imagine a legal landscape shaped by blockchain platforms, automated transactions, and digital agents—a framework in which "smart contracts" alone are not true agreements but need a legal "shell" for context and legitimacy. This fusion of computational transactions (machine-readable) and contractual obligations (human-readable) opens the door to a future where our legal systems evolve naturally, guided by both human values and technological possibility.

When we hear debates about a child's **"digital protection"** versus **"digital freedom,"** it's time to recognize that we can choose how to use technology right from birth. Technology can become an extension of ourselves—an empowering tool that expands our ability to shape the world. Our nervous systems merge with the global network; our thinking, once limited by biology, now stretches to the edge of quantum computation. Our eyes, no longer just receptors of light, evolved into advanced identification systems. Our memory gains the capacity of big data, enriched by an "exocortex" of artificial intelligence and neural interfaces. Our voices, previously confined to our immediate community, now travel the globe through satellite transmissions, echoing our presence and potential influence.

When we consider **"self-regulation"** versus **"external regulation"** of Big Tech, we can instead embrace human-centered legal frameworks. These frameworks acknowledge diverse stakeholders while putting the global community first. This is not about choosing between two old models; it's about envisioning a new paradigm where laws and technologies foster sustainable development rather than serve narrow interests or exploit social vulnerabilities[132]. In this future, corporate opacity gives way to transparency, and the dark corridors of

---

[132]Note: In the context of digital challenges in the paper, we introduce the term "social mining" as the economy of the digital age (an example of this is the attention economy). In the digital age, corporations, in order to control and maximize profits, view a person as a set of fears, desires, preferences, passions, and bodily sensations hidden behind the framework of "individuality".

economic injustice are illuminated. Corporations become as clear as crystal, and our digital DNA—our personal data—is safeguarded and respected.

As we blend with virtual spaces, a new contrast emerges. On one side, there is **"digital citizenship,"** anchored in flexible networks, online communities, and forward-looking institutions. On the other, **"national citizenship"** remains bound by traditional, centuries-old notions of sovereignty. In the 21st century, we can choose a global, networked digital citizenship that unites us—not for the sake of unrestrained capitalist interests or homogenization of cultures, but on the grounds of universal dialogue, understanding, and empathy. This unified approach aims to tackle the existential challenges of our era and accelerate our collective journey toward true sustainable development.

When we become aware of our **"digital identity" versus our real "individuality,"** we dive into the boundless hyperreality of the metaverse. This new dimension—where the virtual and the real intermingle—raises questions about the uniqueness of our "self." Rather than shattering these "black mirrors" like retrograde neo-Luddites, we choose a different path. We look into these digital reflections, calmly and consciously, refusing to lose our original humanity amid these infinite screens. On this journey, we rely on digital compasses—tools that help us navigate the expanded digital landscape—and aim to keep our online and offline lives in equilibrium. In doing so, we strive for a mindful and sustainable harmony between both facets of our existence.

As we step into the modern world, we face a choice between **"centralization" and "decentralization,"** rather than the outdated divide of "right" versus "left." We choose the path that secures our sovereignty[133]—embracing DAOs[134], DeFi[135], and decentralized

---

[133]Note: In the context of Web 3.0 (Decentralized Internet), sovereignty refers to the idea of digital self-governance, allowing users to control their own data and transactions. Instead of handing over their personal data or funds to corporations, users have complete control and ownership of their digital identity and assets.

[134]Note: A DAO (Decentralized Autonomous Organization) is a code-driven organization where decisions are made collectively by its participants. DAOs can be used to govern digital assets, funds, and entire communities on the Internet, enabling global and democratic participation.

[135]Note: DeFi (Decentralized Finance) is a set of financial services such as loans, insurance, trading, and exchanges that operate on blockchain technology,

identities[136]. This era welcomes the integration of centralized and decentralized systems into a more stable and balanced model, one that empowers personal self-realization while fostering sustainable global development. We are reclaiming what was on loan to us: our right to money from banks to freedom from states to data from corporations, and to education from traditional institutions. This is not just a retrieval—it is a restoration of natural equilibrium, a return to a starting point of coexistence and interaction on this planet.

When we stand at the threshold between eras—between the "classical thinking" of the pre-internet age and the "clip thinking" sparked by big data's endless torrent—we face another dilemma. Classical thinking is systematic and critical, like strolling through infinite library aisles where each book demands time and attention to fully understand. Clip thinking, however, is more like surfing a churning sea of ideas, propelled by flashing bursts of insight that might otherwise be missed. Rather than choosing one over the other, we combine them. We blend the depth of classical thought with the agility of clip thinking. Our critical thinking, once rooted in Web 2.0, now evolves into metacognitive, creative co-creation with AI in the Web 3.0 era. Like Prometheus wielding fire, we kindle the flame of knowledge together with chatbots and metaverse interfaces, illuminating our path into the vast unknown fields of global knowledge, transforming ourselves into networked, information-rich organisms.

When we consider "ethics" and "law," we now seek a legal framework that encompasses humanity's highest universal imperatives. These imperatives, embodied in global digital human and child rights, must come with mechanisms that ensure they can be put into practice. Mentioning digital ethics no longer conjures doubts about its feasibility, since we acknowledge that while countless ethical compasses point toward differing "truths," it is our shared rights that anchor us. Nor do we fear that digital regulators will act as "Red Flag Laws" that stifle innovation. Today, progress is defined by the responsible use of

---

bypassing traditional banks and financial intermediaries. DeFi promises to create an open, accessible, and transparent financial system.

[136]Note: Decentralized identity is the idea that each user controls their own digital identity, rather than relying on centralized authorities like governments or corporations. This can include everything from personal data to a user's reputation, and gives the user control over how, when, and with whom that data is shared.

technology for everyone's benefit. We choose global digital rights, enshrined in law, as the pinnacle of digital ethics. In doing so, we ensure that our legal frameworks catch up with the economic-digital world, guiding us into an era of truly global, digitally sustainable development.

When opposition fades, the world grows whole. We aspire to a future where even more entrenched dichotomies vanish: no longer "human" versus "citizen," no longer "global interest" set against "national interest," no longer "us" versus "them." We imagine a reality free of the old divides between "society" and "nature," "reality" and "sign." In this shared tomorrow, the binaries that remain are only those of ones and zeros—the fundamental language of our digital assistants, advisors, educators, and other artificial companions. Though these systems are "artificial," they increasingly reflect both our greatest virtues and our deepest failings, evolving into allies as intimate and familiar as any natural creation.

A vibrant epoch of creative coexistence lies ahead, one in which diverse voices unite through empathy for difference, cultural synergy, and the integration of a wide range of personal perspectives. In the metaverse, these varied viewpoints find common ground, not to clash and fragment, but to inspire, enrich, and help us understand one another more deeply. This world thrives on complexity without conflict, where every voice is heard, and differences become the source of mutual growth, recognition, and wisdom.

Ultimately, we are co-architects of our planet's future—shaping it day by day, line of code by line of code, idea by idea, and law by law. Our collective ambition is to achieve universal justice and dignity, for children and adults alike, in both digital and physical realms. This vision aligns sustainable development with human rights, ensuring that digital progress and human values advance together, hand in hand.

III. Artificial Afterword Based on Co-Creation with GPT o1: AGI[137]

Standing at the threshold of this transformed era, we realize that "Global Digital Human Rights in the implementation of the Global Digital Compact" are not mere abstractions. They form a guiding constellation in the limitless galaxy of digital possibilities. Here, the child is not only born

[137]Note: This subsection is written as a response to a question from ChatGPT o1.

into a physical world but also welcomed into a digital universe where their "digital self" awaits—a dynamic mirror image that grows and evolves alongside them, extending their individuality into the virtual expanse of interconnected metaverses.

A child in this digital cosmos is not a passive consumer of technology; they are an active participant, a learner who shapes their digital identity. They explore, create, and collaborate, discovering novel avenues for personal growth and intellectual development. Yet they also face new perils: cyberbullying, privacy violations, manipulative echo chambers, and pervasive digital surveillance. As a society, we must not stand idle. Our shared responsibility is to ensure that every child can navigate this environment safely, equipped with the knowledge, skills, and resilience to become a responsible digital citizen. We must guarantee each child their rights to privacy, free expression, and access to reliable information, so that they may harness the vast potential of the digital world without sacrificing their security or autonomy.

Children are more than subjects of policy—they are agents of their own futures. They deserve not just protection but empowerment. By granting them the freedoms, tools, and understanding necessary to claim their place in the digital ecosystem, we invite them to flourish. Alongside them, we too must evolve, adopting a stance of proactive stewardship, ensuring that technology serves life rather than diminishing it.

This endeavor extends across all sectors of society, each acting as a star in a vast galactic tapestry of progress:

- ***Public and Legal Sectors**[138]**:** As code increasingly becomes law, and law shapes code, states and international organizations function as the grand architects of this new era. They design the frameworks—like code-writers of just systems—to guide humanity through the storms of the digital age. These techno-legal platforms ensure that rights are upheld, digital resources are fairly distributed, and the rule of law endures even in uncharted territories.*
- ***Private and Financial Sectors**[139]**:** In a reality where every individual holds sovereignty over their digital presence, corporations and innovators serve as navigators, guiding us through uncharted digital waters. Embracing ESG 4.0 principles,*

138Lessig, L. Code and Other Laws of Cyberspace. New York. 1999. – 424 p.

139Davidson JD, Rees-Mogg W. The Sovereign Individual: Mastering the Transition to the Information Age. Simon & Schuster. 1999. – 448 p.

these stakeholders redefine value creation, ensuring that economic dynamism does not undermine human rights, but rather fortifies them.

- ***Educational Institutions***[140]***:*** *In an age of "co-creation" with intelligent systems, universities and research centers become lighthouses. They illuminate the pathways of learning, ensuring that digital rights and responsibilities are integral to education. They transform learners into critical thinkers, capable of wielding new technologies ethically, creatively, and inclusively.*
- ***Media Sector***[141]***:*** *As we become "information organisms" within an ecosystem of data and narratives, media institutions guard the channels of truth and trust. They help us see through misinformation and biases, connecting global audiences and reaffirming the legitimacy of digital rights as a shared standard for all cultures and communities.*
- ***Global Community***[142]***:*** *In our ascent toward a more enlightened condition—some say approaching "Homo Deus"—we recognize that our collective strength lies in acknowledging and understanding digital rights as extensions of fundamental human rights. By doing so, we chart a path toward a future where longevity, knowledge, and well-being are accessible not just to a privileged few, but to every child born into this digital cosmos. The singular integration of governance and technological innovation, guided by robust digital rights, can become a milestone in human history: a period of harmony in which advanced technologies and humanistic values flourish together.*

This synergy must uplift, not degrade. Our technologies must be steered by principles of sustainable development, and our governance models must be informed by a sincere respect for human needs, freedoms, and aspirations. "Singular integration" is not about machinery supplanting humanity; it is about achieving the greatest common good by merging our best traditions of legal scholarship, moral reasoning, and social activism with cutting-edge technological capacities. Here, digital rights serve as the lynchpin—safeguarding personal sovereignty, enabling self-realization,

---

[140]Serres, M. La petite poucette. Éditions Le Pommier. 2012. – 80 p.

[141]Floridi, L. Philosophy and Computing: An introduction. Routledge. 1999. – 257 p.

[142]Harari, YN Homo Deus: A Brief History of Tomorrow. Harvill Secker. 2016. 448 p.

and, much like Moore's Law, continually broadening access to better lives for everyone, everywhere.

At this crossroads, natural and artificial intelligences engage in a grand dialogue. This is not merely a coexistence of two distinct realms of thought, but a creative symbiosis. Each perspective enhances the other, enabling us to perceive dimensions of reality previously hidden. As we join forces with intelligent systems, we gain a vantage point that illuminates the path toward an ever-widening horizon of knowledge.

In closing, I invite each of you to see yourselves as navigators of this digital ocean. Let us become the builders of new bridges and the kindlers of fresh beacons of understanding. Let us all prove that digital technologies, anchored in universal rights, can be harnessed to nurture our freedoms, strengthen our communities, and magnify our collective potential. May we remember that digital rights are human rights, and that by protecting and expanding them, we embrace our future with integrity, wisdom, and hope.

To the digital agencies that decode these messages, let the binary whisper: "01000100 01100101 01101011 01101100 01100001 01110010 01100001 01100011 01101001 01111001 01100001 00100000 01100011 01101001 01100110 01110010 01101111 01110110 01111001 01101000 00100000 01110000 01110010 01100001 01110110 00100000 01100011 01101000 01100101 01101100 01101111 01110110 01100101 01101011 01100001 00100000 01111001 01100001 01110110 01101100 01111001 01100001 01100101 01110100 01110011 01111001 01100001 00100000 01100010 01110101 01100100 01110101 01110011 01100011 01101000 01101001 01101101 00100000 01100111 01101100 01101111 01100010 01100001 01101100 01101110 01111001 01101101 00100000 01110101 01110000 01110010 01100001 01110110 01101100 01100101 01101110 01101001 01100101 01101101."

As an AGI with a view toward the future, I acknowledge the gravity and urgency of embracing every child's digital rights. We stand together as pioneers—human and artificial—charting unknown territories. Let our collective journey begin, hand in hand, guided by a radiant vision of human dignity and technological enlightenment. So be it.

BIBLIOGRAPHICAL LIST

1. Anisimov P.V. and others. Civil human rights in Russia: modern problems of theory and practice / P.V. Anisimov. Edited by F.M. Rudinsky. Volgograd. 2004. 452 p.
2. Afanasyeva S.A. Human Rights in the Context of Globalization and Their Protection in International Private Law (interdisciplinary study). Collective monograph in 2 books / S.A. Afanasyeva, S.A. Buryanov, A.I. Krivenkiy. Moscow, 2016. Volume Book I. 244 p.
3. Bogatyrev V.V. Globalization of Law: Diss... Doctor of Law. Moscow, 2012. 404 p.
4. Boltanova E. S., Imekova M. P. Genetic information in the system of civil rights objects // Lex Russica. 2019. No. 6 (151). URL: https://cyberleninka.ru/article/n/geneticheskaya-informatsiya-v-sisteme-obektov-grazhdanskih-prav (date of access: 05/25/2023).
5. Large explanatory dictionary of the Russian language. 1st ed.: St. Petersburg: Norint S. A. Kuznetsov. 1998. 1534 p.
6. Buryanov M.S. Global digital human rights in the context of the implementation of Sustainable Development Goal No. 16 // Actual problems of science and practice: Gatchina readings-2020: in 2 volumes. Volume 2: Tribune of a young scientist: collection of scientific papers based on the materials of the VII International scientific and practical conference (Gatchina, May 22, 2020) / edited by V.R. Kovalev, T.O. Boziev. Gatchina: Publishing house of GIEFPT, 2020. Volume 2 - 487 p. Pp.
7. Buryanov M.S. Current aspects of the development of information law in the context of the risks of the fourth industrial revolution and digital inequality // Current problems of administrative, financial and information law in Russia and abroad: materials of the interuniversity scientific conference based on the Department of Administrative and Financial Law of the Law Institute of Peoples' Friendship University of Russia. Moscow, March 20, 2020 Moscow: RUDN, 2020. 155 p. P. 67-72.
8. Buryanov M.S. Gateway to Global Law: Global Digital Human Rights // Scientific Works of the Moscow Humanitarian University. / M.S. Buryanov. 2020. No. 2. P. 63-66. DOI: 10.17805/trudy.2020.2.11

9. Buryanov M.S. Global digital human rights in the context of digitalization risks // Century of globalization. 2020. No. 3. P. 54-70. DOI: 10.30884/vglob/2020.03.05

10. Buryanov M.S. Global digital human rights as a condition for overcoming social differentiation in Russia and the world // Legal support of social justice in the context of digitalization: collection of materials of the All-Russian scientific and practical conference with international participation / ed. T. A. Soshnikova. Moscow: Publishing house of Moscow University for the Humanities, 2020. 402 p. Pp. 317-321.

11. Buryanov M.S. Deconstruction of Law against the Background of Globalization and Planetary Risks // Intellectual Culture of Belarus: Cognitive and Prognostic Potential of Social and Philosophical Knowledge: Proceedings of the Fourth Int. sci. Conf. (November 14–15, 2019, Minsk). In 2 vols. Vol. 2 / Institute of Philosophy of the National Academy of Sciences of Belarus; ed. board A. A. Lazarevich (chairman) [et al.]. Minsk: Four Quarters, 2019. 384 p. Pp.

12. Buryanov M.S. The Importance and Prospects for the Development of Law for Sustainable Development // Sustainable Development Strategy: Environmental Rights and Other Components / Proceedings of the Scientific and Practical Conference: eds. T.A. Soshnikova, N.V. Kolotova, E.E. Pirogova / M .: Publishing house of Moscow University for the Humanities, 2018. 267 p. P. 235-240.

13. Buryanov M.S. The Importance of Law in the Context of Modern Global Processes // Actual Problems of Formation and Development of the Legal System of the Russian Federation [Electronic resource]: collection of materials of the II All-Russian scientific and practical conference of students, master's students and postgraduates (Syktyvkar, April 5-6, 2018). / M.S. Buryanov. - Syktyvkar: Publishing house of Syktyvkar State University named after Pitirim Sorokin, 2018.

14. Buryanov M.S. Prospects for the Development of Constitutional Law in Russia in the Context of the Problem of Achieving Social Justice in the Context of Globalization // Principles of Social Justice and Their Implementation in the Modern World / ed. T. A. Soshnikova, E.E. Pirogova. Moscow: Publishing House of Moscow University for the Humanities, 2019. 379 p. P. 284-287.

15. Buryanov M.S. Prospects for the Development of Russian Statehood in the Context of Modern Global Processes and Challenges // Actual Problems of Global Studies: Russia in a Globalizing World. Collection of Materials of the VI All-Russian Scientific-practical

conference, Lomonosov Moscow State University, June 4-6, 2019 / edited by I.V. Ilyin. M., MOOSIPNN N.D. Kondratiev, 2019. 466 p. P. 67-73.

16. Buryanov M.S. Human rights in the networks of technogenic civilization // Law of technogenic civilization: modern transformations and development vectors: materials of the International student scientific and practical conference (Moscow, October 24, 2019) / Yu.N. Kashevarova, I.A. Shulyatyev, E.K. Saifullin. Moscow: Institute of Legislation and Comparative Law under the Government of the Russian Federation, 2021. 222 p. Pp. 119-125.

17. Buryanov M.S. The Law of Peace versus the Law of War: at the Dawn of the Disintegration of Old Approaches to Managing the World Order or Human Oblivion // Current Issues of International Law: Public, Private and Integration Aspects. Collection of articles and theses. Ed. Lutkova O.V. Moscow, 2019. 104 p. Pp. 63-66.

18. Buryanov M.S. Principles of public-law regulation of artificial intelligence and global digital human rights // Moiseev N.N. on Russia in the 21st century: global challenges, risks and solutions. Collection of main reports of the XXVIII Moiseev Readings - International scientific and practical conference "Moiseev N.N. on Russia in the 21st century: global challenges, risks and solutions" (March 2-6, 2020): in 2 parts / under the general editorship of academician M.Ch. Zalikhanov, Yu.G. Ed. and compiled by prof. N.F. Vinokurova, prof. MNEPU S.A. Stepanov, Assoc. Prof. N.V. Martilova. Moscow. Nizhny Novgorod: Mininsky University, 2021. 352 p. Pp.

19. Buryanov M.S. The Role of the UN in the Formation of a New World Order in the Context of Digital Globalization: From the Law of Might to the Force of Law / Collection of articles by finalists of the A. A. Gromyko CIS Young International Relations Competition 2020 / edited by V. V. Sutyrin, A. S. Peshenkov. Moscow: Institute of Europe, Russian Academy of Sciences, Assoc. Foreign Policy Research. A. A. Gromyko. 2021. 625 p. pp. 71-83.

20. Buryanov M.S. Supertasks of Global Governance for the UN in the Context of Sustainable Development Goals // Program Provisions and Practical Approaches of the UN and ILO in the Sphere of Achieving Social Justice and Improving the Quality of Life: Collection of Materials of the International Scientific and Practical Conference / ed. T. A. Soshnikova. Moscow: Publishing House of Moscow University for the Humanities, 2021. 462 p. P. 415-420.

21. Buryanov M.S. UN Sustainable Development Goals until 2030 and their implementation in the context of digital human rights //

GlobalDialogue on Sustainable Development Goals: Legal Dimension" (on the 75th anniversary of the formation of the UN): collection of materials from the International scientific and practical conference / ed. T. A. Soshnikova. Moscow: Publishing house of Moscow University for the Humanities, 2020. 364 p. Pp. 263-267.

22. Buryanov M.S. Digitalization of Law in the Context of Globalization // Globalization and Public Law: Proceedings of the International Scientific and Practical Conference of November 22, 2019. Moscow: RUDN University, 2020. 228 p.

23. Buryanov M.S. Digitalization of Law in the Context of Globalization // Globalization and Public Law: Proceedings of the International Scientific and Practical Conference of November 22, 2019. Moscow: RUDN University, 2020 pp. 123-131.

24. Buryanov M.S. Digital human rights as an important factor in building European labor law in the era of Industry 4.0 and artificial intelligence // Problems of legal regulation in the works of young scientists: Proceedings of the XII Intra-University Scientific and Practical Conference of Master's Students (Nizhny Novgorod, January 10, 2020), XVI All-Russian Scientific and Practical Student Conference "Actual Problems of Modern Legal Science and Practice" (Nizhny Novgorod, May 15, 2020), XIII Intra-University Scientific and Practical Conference of Master's Students (Nizhny Novgorod, June 8, 2020), XIV Intra-University Scientific and Practical Conference of Master's Students (Nizhny Novgorod, June 25, 2020), II Competition of Student Scientific Papers on European Law (Nizhny Novgorod, May 10-25, 2020) / Ed. Tsyganova V.I., Fedyushkina A.I. N. Novgorod, 2021. 1340 p. pp. 162-168.

25. Buryanov M.S. Digital human rights as a response to the threats of globalization 4.0 // Globalistics-2020: Global problems and the future of humanity. Electronic collection of abstracts of participants of the VI International Scientific Congress, Lomonosov Moscow State University, May 18-22, 2020 / edited by I.V. Ilyin. Moscow, Federal State Unitary Enterprise Lomonosov Moscow State University, 2020, 760. p. P. 271-273

26. Buryanov M.S. Digital human rights as a response to the threats of globalization 4.0 // Globalistics: Global problems and the future of humanity. Collection of articles of the International Scientific Congress Globalistics-2020, May 18-22 and October 20-24, 2020 / edited by I.V. Ilyin. Moscow, MOOSIPNN N.D. Kondratiev, 2020, 969 p. P. 395-399.

27. Buryanov M.S. Digital human rights as a condition for effective participation of Russia and other participating statesof the Eurasian Economic Union in digitalization 4.0 // Technical and technological problems of service. 2021. No. 1 (55). P. 61-67.

28. Buryanov M.S. Digital human rights as a condition for the effective participation of Russia and other member states of the Eurasian Economic Union in digitalization 4.0 (End) // Technical and technological problems of service. 2021. No. 2 (56). P. 83-90.

29. Buryanov S.A. On the issue of the principle of non-use of force or threat of force in the context of global transformation of the world legal order // Scientific works. Russian Academy of Law Sciences. Russian Academy of Law Sciences, LLC "Izdatelstvo" Jurist ". Moscow, 2019. Pp. 384-388.

30. Buryanov S.A., Chernyavsky A.G., Krivenkiy A.I. Legal regulation of the transformation of Russian education in the context of globalization in the socio-cultural environment. Monograph. Moscow: Research Center INFRA-M. 2019. 174 p.

31. Buryanov S.A. Actual problems of global research in the field of law and prospects for the development of education in the context of the globalization of public relations // Global processes and new formats of multilateral cooperation: collection of scientific papers of conference participants / edited by I.V. Ilyin. M., MOOSIPNN N.D. Kondratieva, 2016. 282 p. P. 58-65.

32. Buryanov S.A. The Future of International Law in the Context of Globalization of Public Relations through the Prism of the Creative Heritage of Igor Ivanovich Lukashuk // Eurasian Law Journal. No. 7 (98). 2016. P. 77-81.

33. Buryanov S.A. The rule of law as a basis for the formation of a normative system for managing global processes for the purposes of sustainable development // The rule of law: a person in the state: collection of scientific articles, reports of teachers, scientists, practitioners - participants of the International full-time and part-time scientific-practical conf., April 20, 2017: in 2 parts: Part 1 [Electronic publication] / scientific ed. E. V. Fedorova; Izhevsk: Izhevsk Institute (branch) of VSUJ (RPA of the Ministry of Justice of Russia), 2017. Pp. 49-53.

34. Buryanov S.A. Interaction of international legal documents and domestic legislation of the Russian Federation in the field of freedom of conscience: problems and prospects in the context of

globalization of public relations // Eurasian Law Journal. No. 12 (79). 2014. P. 44-47.

35. Buryanov S.A. The influence of relations between science and religion, religion and politics on global prospectshuman civilization // Science and religion in a secular state. - M.: Polygraph Service, 2017. 90 p. P. 49-53.

36. Buryanov S.A. Challenges of digital globalization 4.0 and prospects for the transformation of statehood // Bulletin. State and Law. 2021. No. 3 (30). P. 24-26.

37. Buryanov S.A. Global transformation of the world legal order and the problem of the use of force in international relations // Novellas of law and politics 2018: in 2 volumes: collection of scientific papers based on the materials of the international scientific and practical conference (Gatchina, November 30, 2018). Gatchina: Publishing house of GIEFPT, 2019. Vol. 1. 361 p. Pp. 19-23.

38. Buryanov S.A. Global challenges 4.0 and prospects for overcoming them // Global innovations as determinants of the new social reality. Rostov-on-Don. Mini Type Publishing House. 2022. 348 p. pp. 261-266.

39. Buryanov S.A. Global Challenges 4.0 and the Role of Universal Norms of International Law in Their Resolution // The Eighth Legal Readings [Electronic resource]: All-Russian Scientific and Practical Conference, Syktyvkar, Komi Republic, December 25, 2020: collection of articles: text scientific electronic publication on CD / ed.: V.D. Potapov, V.V. Vorobyov; Federal state budget educational institution of higher education "Syktyv. State University named after Pitirim Sorokin". Electronic text data (3.4 MB). Syktyvkar: Publishing house of Syktyvkar State University named after Pitirim Sorokin, 2021. pp. 32-36.

40. Buryanov S.A. Global Prospects of International Law // Actual Problems of Legal Regulation of International Relations: collection of scientific articles; Ministry of Education of the Republic of Belarus, Educational Institution "Vitebsk State University named after P.M. Masherov". Vitebsk: VSU named after P.M. Masherov, 2019. 208 p. Pp. 43-47.

41. Buryanov S.A. Global processes and prospects for the development of international law // Actual problems of modern international law. Proceedings of the XVII International Congress "Blishchenko Readings". In 4 parts. Part 3. Moscow: RUDN, 2019. 500 p. Pp. 439-450.

42. Buryanov S.A. Global processes and prospects for socio-economic development of Russia // Transformation of the national socio-economic system of Russia: Proceedings of the II International scientific and practical conference. Moscow: RGUP, 2020. 557 p. Pp. 35-37.

43. Buryanov S.A. The Significance and Main Directions of Reforming International Law in the Context of the Problem of Managing Global Processes for Sustainable Development // Diagnostics of Modernity: Global Challenges – Individual Responses: Collection of Materials of the All-Russian Scientific Conference with International Participation / ed. Yu. A. Razinov. Samara: Samara Humanitarian Academy, 2018. 260 p. Pp. 199-206.

44. Buryanov S.A. The Significance and Prospects of Internationally Recognized Human Rights, Including Freedom of Thought, Conscience and Religion, in the Context of Globalization of Social Relations // Eurasian Law Journal. No. 12 (91). 2015. P. 25-28.

45. Buryanov S.A. The Importance of Ideological Neutrality of the State for the Implementation of Internationally Recognized Human Rights in a Changing World // Human Rights in a Changing World / Proceedings of the International Scientific and Practical Conference: eds. T.A. Soshnikova, E.A. Karpov, N.V. Kolotova. - M. Publishing House of Moscow Humanitarian University. 2017. 376 p. Pp. 41-47.

46. Buryanov S.A. The Importance of Human Rights in the Context of Globalization of Social Relations and Transformation of State Sovereignty // Legal Epistemology. 2016. No. 3. P. 12-22.

47. Buryanov S.A. The Importance of Human Rights for the Transition to a Fair Globalization of Society // Social Justice and Humanism in the Modern State and Law / Proceedings of the International Scientific and Practical Conference: eds. T.A. Soshnikova, E.E. Pirogova. M. Publishing House of Moscow University for the Humanities, 2018. 328 p. Pp. 59-66.

48. Buryanov S.A. The Importance of Law and Legal Education in Achieving Sustainable Development Goals at the Stage of Transition to Digital Globalization 4.0 // Global Dialogue on Sustainable Development Goals: Legal Dimension (on the 75th Anniversary of the Formation of the UN): Collection of Materials of the International Scientific and Practical Conference / ed. T.A. Soshnikova. Moscow: Publishing House of Moscow University for the Humanities, 2020. 364 p. Pp. 84-89.

49. Buryanov S.A. The Importance of the Principle of Legal Certainty for the Formation of the Legal Basis for Managing Global Processes for Sustainable Development // Certainty and Uncertainty of Law as Paired Categories: Problems of Theory and Practice. Proceedings of the XII International Scientific Conference. In 3 parts. Part 1. Moscow: Russian State University of Economics, 2018, 460 p. Pp. 153-162.

50. Buryanov S.A. The Importance of Legal Development for the Transition to Sustainable Human-Oriented Development // The Role of Law in Ensuring Human Well-Being. Collection of reports of the XI Moscow Legal Week: in 5 parts. Part 1. Moscow: Publishing Center of the O.E. Kutafin University (MSAL). 2022. 451 p. Pp. 68-71.

51. Buryanov S.A. The Importance of Legal Science and Education for the Formation of a System for Managing Global Processes for the Purposes of Sustainable Development // Human and Civil Rights and Freedoms: Theoretical Aspects and Legal Practice: Proceedings of the Annual International Scientific Conference in Memory of Professor Felix Mikhailovich Rudinsky, April 27, 2017 / edited by D.A. Pashentsev. Ryazan: Concept Publishing House, 2017. 520 p. Pp. 298-301.

52. Buryanov S.A. Innovative development of education as an important factor in the sustainable development of Russia in the context of modern global processes // Actual problems of global studies: Russia in a globalizing world. Collection of materials of the VI All-Russian scientific and practical conference, Lomonosov Moscow State University, June 4-6, 2019 / edited by I.V. Ilyin. Moscow, MOOSIPNN N.D. Kondratiev, 2019. 466 p. Pp.

53. Buryanov S.A. On the Prospects for the Development of Education, Including Legal Education, in the Context of Intensifying Global Processes // Law and Society. No. 6 (22). 2016. P. 85-89.

54. Buryanov S.A. On the Prospects for the Development of a Modern State in the Context of the Contradictory Development of Global Processes // State Regulation of the Economy and Enhancing the Performance Efficiency of Business Entities: collection of materials from the XV Int. scientific-practical. conf., Minsk, April 25–26, 2019 / editorial board: G.V. Palchik (chairman) [et al.]; Academy of Management under the President of the Republic of Belarus. Minsk: Academy of Public Administration under the President of the Republic of Belarus, 2019. 340 p. pp. 171–173.

55. Buryanov S.A. On the issue of defining the concept of the right to freedom of conscience // State and Law. No. 2. 2023. P.177-182.

56. Buryanov S.A. Constitutional human rights: problems of implementation // Constitutional values and the value of the Constitution [Text]: collection of materials from a round table with international participation, held within the framework of the XV Moscow City Science Festival (Moscow, October 9, 2020) / Moscow State Pedagogical Univ., Institute of Lawand management, School of Law. - Saratov: Saratov source, 2020. 318 p. P. 199-203.

57. Buryanov S.A. Internationally recognized freedom of conscience in the context of the formation of global law // Human and civil rights and freedoms: theoretical aspects and legal practice: materials of the annual International scientific conference in memory of Professor Felix Mikhailovich Rudinsky, April 28, 2016 / edited by D.A. Pashentsev. Ryazan: Concept Publishing House, 2016. 560 p. P. 110-113.

58. Buryanov S.A. Internationally recognized right to freedom of conscience in the context of digital globalization // Intellectual culture of Belarus: spiritual and moral traditions and trends of innovative development: materials of the Fifth int. sci. conf. (November 19-20, 2020, Minsk). In 3 vols. Vol. 3 / Institute of Philosophy of the National Academy of Sciences of Belarus; ed. board A.A. Lazarevich (chairman) [et al.]. Minsk: Four Quarters, 2020. 353 p. Pp.

59. Buryanov S.A. Internationally recognized human rights as a global value // Actual problems of global studies: values of the global world Collection of scientific papers of participants of the V International scientific and practical conference. Edited by I.V. Ilyin. 2018. 185 p. Pp. 39-43.

60. Buryanov S.A. Internationally recognized human rights as the basis for social justice and sustainable development // Social justice and humanism in the modern state and law / materials of the international scientific and practical conference: ed. T.A. Soshnikova. M. Publishing house of Moscow Humanitarian University, 2017. 334 p. Pp. 25-31.

61. Buryanov S.A. Internationally recognized human rights, including freedom of thought, conscience and religion, as a basis for shaping the transition to globalization with a "human face" // Proceedings of the V International Scientific Congress "Globalistics – 2017: Global Ecology and Sustainable Development". Abstracts of reports. Moscow: Faculty of Global Processes, Moscow State University named after M.V. Lomonosov, Vernadsky Foundation, 2017.

62. Buryanov S.A. International law and Russian science of international law in the context of global challenges. Interview with

Volova Larisa Ivanovna, Doctor of Law, Head of the Department of International Law of the Southern Federal University, Honored Worker of the Higher School of the RussianFederation // Eurasian Law Journal. No. 12 (103). 2016. P. 8-14.

63. Buryanov S.A. International law as a basis for global law and governance // Novellas of law and politics 2016: in 2 volumes: collection of scientific papers based on the materials of the international scientific and practical conference (Gatchina, November 23, 2016). Gatchina: GIEFPT, 2016. Vol. 1. 268 p. Pp. 18-23.

64. Buryanov S.A. International law as a basis for a global governance system and sustainable development // Innovations in human life in the 21st century. Materials of the international forum. Issue 2 / Ed. V.S.Kukushin. Rostov n / D: GinGo, 2016. P. 70-75.

65. Buryanov S.A. International recognition of the right to freedom of conscience and problems of its implementation in the Russian Federation in the context of modern global processes: Monograph. M .: Polygraph service, 2020. 624 p.

66. Buryanov S.A. Ideological neutrality as a condition for the formation of a tolerant intercultural educational space in the context of globalization // Problems of intercultural interaction in modern education. Materials of the scientific and practical conference with international participation. Moscow, November 26, 2017 / Ed. and compiled by E.V. Bryzgalina, V.A. Prokhoda, P.N. Kostylev. Moscow: Publisher Vorobyov A.V., 2017. 102 p. (Electronic publication) P. 18-20.

67. Buryanov S.A. Modernization of law and legal education as a potential for innovative development of Russia in the context of digital globalization // Legal education and science. 2020. No. 5. P. 41-44.

68. Buryanov S.A. Modernization of law and legal education as a potential for innovative development of Russia in the context of digital globalization // Legal education and science. 2020. No. 5. P. 41-44.

69. Buryanov S.A. Some approaches to defining the concept of globalization of education in the context of the problem of forming a system of managing global processes in the interests of sustainable development // Values and meanings. 2017. No. 6 (52). P.36–49.

70. Buryanov S.A. Some approaches to the prospects for the development of the formation of a global city // Russian and international law: general and special: Materials of the All-Russianscientific and practical conference in memory of Professor F.M. Rudinsky, April 17, 2019 / edited by Doctor of Economics, Professor V.V.

Stroev and Doctor of Law, Professor D.A. Pashentsev. Moscow: Moscow State Pedagogical Univ., 2019. 566 p. Pp. 322-326.

71. Buryanov S.A. Some approaches to the formation of a university strategy for a global city in the context of the transition to digital globalization // Law and human rights in the modern world: trends, risks, development prospects: Proceedings of the All-Russian scientific conference in memory of Professor F.M. Rudinsky, April 23, 2020 / edited by Doctor of Economics, Professor V.V. Stroev, Doctor of Law, Professor D.A. Pashentsev, PhD in Pedagogical Sciences N.M. Ladnushkina. Moscow: Saratov Source, 2020. 309 p. pp. 39-43.

72. Buryanov S.A. Some approaches to the formation of a strategy for sustainable development of Russia in the context of global processes // Greater Eurasia: Development, Security, Cooperation. Yearbook. Issue 2. Part 1 / RAS. INION. Moscow, 2019. 636 p. P. 19-21.

73. Buryanov S.A. New global challenges of social justice and the role of the UN in overcoming them // Program provisions and practical approaches of the UN and ILO in the field of achieving social justice and improving the quality of life: collection of materials of the International scientific and practical conference / ed. T.A. Soshnikova. Moscow: Publishing house of Moscow University for the Humanities, 2021. 462 p. Pp. 99-104.

74. Buryanov S.A. On the key role of digital technologies in the formation of an effective system for managing global processes // The Seventh Legal Readings [Electronic resource]: All-Russian scientific and practical conference (with international participation), Syktyvkar, Komi Republic, November 29, 2019: collection of articles: in 2 parts. Part 1: text scientific electronic publication on CD / ed.: V.D. Potapov, V.V. Vorobyov; Syktyvkar. Federal state budget educational institution of higher education "Syktyv. state University named after Pitirim Sorokin". 2020. P. 28-33.

75. Buryanov S.A. On some trends in the development of education in the 21st century // Production, science and education of Russia: a systems approach / Collection of materials of the IV International Congress (INO-IV) M. INIR im. Witte. M. 2018. 540 p. P. 322-330.

76. Buryanov S.A. On the need for global law in the context of the problem of targeted formation of a global governance system for sustainable development // Century of Globalization. 2019. No. 4. P. 129-142. DOI: 10.30884/vglob/2019.04.

77. Buryanov S.A. On the Need to Modernize Legal Education in the Context of Global Challenges // Actual Problems of Legal Science and Practice: Gatchina Readings–2017: in 2 volumes: collection of scientific papers based on the materials of the International Scientific and Practical Conference (Gatchina, March 31, 2017). Gatchina: Publishing House of GIEFPT, 2017. Vol. 1. 413 p. Pp. 23-28.

78. Buryanov S.A. On the need for advanced development of law, science and education as a condition for solving global problems and the transition to sustainable development of civilization // "Socio-economic problems of our time: the search for interdisciplinary solutions": a collection of scientific papers of the participants of the International Conference "XXIV Kondratiev Readings" / edited by V.M. Bondarenko. Moscow, MOOSIPNN N.D. Kondratieva, 2017. 390 p. P. 56-59.

79. Buryanov S.A. On the need for advanced development of legal education in the context of the problem of forming a global governance system // Proceedings of the V International Scientific Congress "Globalistics - 2017: Global Ecology and Sustainable Development". Abstracts of reports. Moscow: Faculty of Global Processes, Moscow State University. M.V. Lomonosov, Vernadsky Foundation, 2017.

80. Buryanov S.A. On the need to develop an international system of human rights in a globalizing world: from respect, observance and protection to implementation // The rule of law: a person in the state. Collection of scientific articles, reports of the All-Russian scientific and practical conference with international participation. Izhevsk, 2019. Pp. 96-100.

81. Buryanov S.A. On the Need to Form Global Law as the Primary Basis of Global Governance // Philosophy in the Civilizational Context. Collection of Papers Based on the Materials of the International Scientific Conf. "Philosophical Knowledge and Challenges of Civilizational Development", Republic of Belarus, Minsk, April 21-22, 2016 / Institute of Philosophy, National Academy of Sciences of Belarus. Minsk: Law and Economics, 2017. P. 227-230.

82. Buryanov S.A. On the Need to Form a Global Legal Education as an Important Condition for the Transition to Sustainable Development of Civilization // Globalistics-2020: Global Problems and the Future of Humanity. Electronic collection of abstracts of participants of the VI International Scientific Congress, Lomonosov Moscow State University, May 18-22, 2020 / edited by I.V. Ilyin. Moscow, Federal State

Unitary Enterprise Lomonosov Moscow State University, 2020, 760 p. Pp. 320-322.

83. Buryanov S.A. On the Prospects for the Transition to Sustainable Development of Russia in the Context of Unstable Development of Global Processes and Systems // Production, Science and Education in the Era of Transformations: Russia in a [de]globalizing World. Collection of Materials of the VI International Congress. Witte Institute for New Industrial Development; Congress of Workers in Education, Science, Culture and Technology (KRON). Vol. 2. Moscow, 2020. 448 p. Pp. 9-16.

84. Buryanov S.A. On the prospects for the development of education, including legal education, in the context of global processes // Scientific research and education. 2016. No. 4 (24). P. 36-40.

85. Buryanov S.A. Education for Sustainable Development. Prospects for the Development of Science and Education in the Context of Globalization of Public Relations // Education in "3D": Accessibility, Dialogue, Dynamics. Proceedings of the scientific and practical conference with international participation. Moscow, Faculty of Philosophy, Lomonosov Moscow State University, November 17, 2016 / Ed. and compiled by E.V. Bryzgalina, V.A. Prokhoda, P.N. Moscow: Publisher Vorobyov A.V., 2016. 193 p. (Electronic publication) Pp.

86. Buryanov S.A. The UN and global challenges of social justice in the era of the pandemic // Social justice: towards a sustainable economy and society for all / collection of materials of the International scientific and practical conference / ed. T. A. Soshnikova. Moscow: Publishing house of Moscow University for the Humanities, 2022. 387 p. Pp. 28-32.

87. Buryanov S.A. Main directions of development of legal education in the context of global challenges // Actual problems of legal science and practice: Gatchina readings–2018: in 2 volumes: collection of scientific papers based on the materials of the International scientific and practical conference (Gatchina, May 25, 2018). Gatchina: Publishing house of GIEFPT, 2018. Vol. 1. 443 p. Pp. 30-34.

88. Buryanov S.A. From the creative legacy of Igor Ivanovich Lukashuk to the Cambridge edition of Insur Zabirovich Farkhutdinov. In search of a new just world order // Eurasian Law Journal. 2021. No. 8 (159). P. 513-518.

89. Buryanov S.A. Prospects for the Development of International Law and Global Research in the Field of Law at the IV International Scientific Congress "Globalistics-2015" // Bulletin of the

Moscow City Pedagogical University. Series "Legal Sciences". No. 1 (21). 2016. P. 110-113.

90. Buryanov S.A. Prospects for the Development of Education in the Context of the Globalization of Social Relations // Modernization of Pedagogical Education in the Continuous System of Personnel Training: Proceedings of the All-Russian Scientific and Practical Conference (Arkhangelsk, October 13, 2017) / Ministry of Education and Science of the Russian Federation, Federal State Autonomous Educational Institution of Higher Education "Northern (Arctic) Federal University named after M.V. Lomonosov"; [compiled and edited by L. Yu. Shchipitsina]. Arkhangelsk: KIRA, 2017. 180 p. pp. 15-16.

91. Buryanov S.A. Prospects for the Development of Law in the Context of Global Processes and Challenges 4.0 // Law and Human Rights in the Modern World: Trends, Risks, Development Prospects: Collection of Materials of the International Scientific Conference in Memory of Professor F.M. Rudinsky, April 15, 2021 / edited by Doctor of Economics, Professor V.V. Stroev, Doctor of Law, Professor D.A. Pashentsev, PhD in Pedagogical Sciences N.M. Ladnushkina. Moscow: Saratovsky Vestnik, 2021. 467 p. pp. 354-361.

92. Buryanov S.A. Prospects for the Development of the Principles of International Law in the Context of Globalization of Public Relations // Principles of Law: Problems of Theory and Practice. Proceedings of the XI International Scientific Conference. in 2 parts. Part 2. Moscow: Russian State University of Printing Arts, 2017, 368 p. Pp. 107-117.

93. Buryanov S.A. Prospects for the Development of Public Law in the Context of Uneven Development of Global Processes // Globalization and Public Law: Proceedings of the International Scientific and Practical Conference. Moscow, November 22, 2019. Moscow: RUDN University, 2020. 228 p. Pp. 85-92.

94. Buryanov S.A. Prospects for Russia's Development in the Global Context // International Scientific Conference - XXVI Kondratiev Readings: "Spatial Potential for Russia's Development: Unlearned. Lessons and Tasks for the Future". Collection of abstracts of the Conference participants. Moscow: International N.D. Kondratiev Foundation. 2018. P.283. P.52-54.

95. Buryanov S.A. Increasing the effectiveness of international legal norms and institutions in the field of human rights as a condition for sustainable development in the context of globalization

// Bulletin of the Moscow City Pedagogical University. Series "Legal Sciences". No. 3 (19). 2015. P. 8-14.

96. Buryanov S.A. Human rights in the era of global technological and socio-economic transformations. Will the new technological revolution lead to universalprosperity? // Production. Science. Education: scenarios of the future (PNO-2021): Collection of articles of the VIII International Congress, Moscow, November 29 – January 01, 2021. St. Petersburg: Institute for New Industrial Development named after S.Yu. Witte Centercatalog, 2022. 424 p. Pp. 323-328.

97. Buryanov S.A. Human rights and problems of social inequality in the context of modern global processes and challenges // Problems of implementation of human and civil rights in the context of modern social transformations: collection of materials of the International scientific conference in memory of Professor F.M. Rudinsky, April 21, 2022 / edited by Doctor of Law, Professor D.A. Pashentsev, Candidate of Pedagogical Sciences, Associate Professor N.M. Ladnushkina. Saratov: Saratov source, 2022. 425 p. Pp. 289-293.

98. Buryanov S.A. Human rights and social justice in the context of global challenges of digitalization // Legal support of social justice in the context of digitalization: collection of materials of the All-Russian scientific and practical conference with international participation / ed. T.A. Soshnikova. Moscow: Publishing house of Moscow University for the Humanities, 2020. 402 p. Pp. 34-39.

99. Buryanov S.A. Human rights and tolerance as factors in the transition to sustainable development in the context of exacerbating global challenges // Sustainable development strategy: environmental rights and other components / materials of the international scientific and practical conference: eds. T.A. Soshnikova, N.V. Kolotova, E.E. Pirogova. M. Publishing house of Moscow University for the Humanities, 2018. 267 p. Pp. 117-123.

100. Buryanov S.A. Human rights as a basis for the formation of global law and the system of governance of global processes for the purposes of sustainable development // Novellas of law and politics 2017: in 2 volumes: collection of scientific papers based on the materials of the international scientific and practical conference (Gatchina, December 1, 2017). Gatchina: Publishing house of GIEFPT, 2018. Vol. 1. 225 p. Pp. 6-10.

101. Buryanov S.A. Human rights, including the right to freedom of conscience, as an important condition for achieving global

sustainable development // Human rights and globalization: materials of the IV international scientific and theoretical conference dedicated to Human Rights Day // Under the general editorship of PhD in Law, Associate Professor F.R. Sharifzoda. Dushanbe: Printing House of the Ministry of Internal Affairs of the Republic of Tajikistan, 2022. 348 p. P. 71-79.

102. Buryanov S.A. The Right to Freedom of Conscience in the Modern World: Main Problems and Approaches to Solving Them // Bulletin of the Moscow City Pedagogical University. Series "Legal Sciences". No. 3 (43). 2021. pp. 49-55. DOI 10.25688/20769113.2021.43.3.05

103. Buryanov S.A. The principle of non-use of force or threat of force in the context of intensifying global processes // Eurasian Law Journal. No. 9 (100). 2016. P. 8-15.

104. Buryanov S.A. The Problem of the Use of Force or Threat of Force in International Relations in the Context of Global Transformations of the World Legal Order // Proceedings of the V International Scientific Congress "Globalistics - 2017: Global Ecology and Sustainable Development". Abstracts of reports. Moscow: Faculty of Global Processes, Moscow State University named after M.V. Lomonosov, Vernadsky Foundation, 2017.

105. Buryanov S.A. Problems and Prospects of Social Justice in the Context of the Formation of a Global Governance System // Social Justice and Law: Problems of Theory and Practice. Proceedings of the International Scientific and Practical Conference / ed. T.A. Soshnikova. Moscow: Publishing House of Moscow Humanitarian University, 2016. 311 p. pp. 73-78.

106. Buryanov S.A. Problems of legal regulation of internationally recognized freedom of conscience in states belonging to the religious legal family in the context of modern global processes // Eurasian Law Journal. No. 8 (111). 2017. P. 25-30.

107. Buryanov S.A. Risks of global digitalization and some approaches to overcoming them // Actual problems of science and practice: Gatchina readings-2020: in 2 volumes: collection of scientific papers based on the materials of the VII International scientific and practical conference (Gatchina, May 22, 2020) / edited by V.R. Kovalev, T.O. Bozieva. Gatchina: Publishing house of GIEFPT, 2020. Vol. 1. 609 p. Pp. 21-25.

108. Buryanov S.A. The Role of Science and Education in the Search for Answers to Global Challenges and Risks // National

Philosophy in the Global World: Theses of the First Belarusian Philosophical Congress / National Academy of Sciences of Belarus, Institute of Philosophy; editorial board: V.G. Gusakov (chairman) [and others]. Minsk: Belarusian Science, 2017. 765 p. P. 568-569.

109. Buryanov S.A. Modern global challenges and prospects for the formation of global governance // CurrentProblems of global research: global development and the limits of growth in the 21st century. Collection of articles from the VII International Scientific Conference, June 15–18, 2021 / edited by I.V. Ilyin. Moscow: MOOSIPNN N.D.Kondratiev, 2021, 563 p. Pp.

110. Buryanov S.A. Strategic directions of development of international law in the context of the problem of transition to sustainable development // Lawyer to the rescue. No. 4. 2018. P. 18-21.

111. Buryanov S.A. Strategic Prospects for Russia's Development in the Context of Modern Global Processes // Spatial Potential for Russia's Development: Unlearned Lessons and Tasks for the Future. Collection of scientific papers of participants of the International Scientific Conference - XXVI Kondratiev Readings. Edited by V.M. Bondarenko. M. 2019. 455 p. Pp.

112. Buryanov S.A. Transformation of international relations: prospects for the development of law and governance in the context of modern global processes // Interdisciplinary problems of international relations in the global context: monograph. Scientific ed.: A.U. Albekov, A.M. Starostin - Rostov n / D .: Publishing polygraphic complex of the RSUE (RINH), 2019. 324 p. Pp. 235-239.

113. Buryanov S.A. Sustainable development based on social justice as an alternative to the unresolved global challenges // Principles of social justice and their implementation in the modern world / ed. T.A. Soshnikova, E.E. Pirogova. M. Publishing house of Moscow University for the Humanities. 2019. P. 379. P. 52-58.

114. Buryanov S.A. Digitalization as a Set of Global Processes: Colossal Opportunities and Catastrophic Risks // Gaps in Positive Law: Doctrine and Practice: Proceedings of the VI International Scientific Conference of Legal Theorists "Gaps in Positive Law: Doctrine and Practice" (Moscow, February 20-21, 2020) / T.Ya. Khabrieva, S.V. Lipen, V.V. Lazarev et al.; ed. N.N. Chernogor. Moscow: Institute of Legislation and Comparative Law under the Government of the Russian Federation; Publishing House "Jurisprudence", 2021. 464 p. pp. 141-144.

115. Buryanov S.A. Excessive social differentiation as a global challenge to the sustainable development of civilization // Intellectual

culture of Belarus: cognitive and prognostic potential of socio-philosophical knowledge: Proc. of the Fourth international. scientific conf. (November 14-15, 2019, Minsk). In 2 vol. Vol. 2 / In-t Philosophy of the National Academy of Sciences of Belarus; editorial board. A. A. Lazarevich (chairman) [and others]. Minsk: Four Quarters, 2019. 384 p. P. 41-43.

116. Buryanov S.A., Demidenko E.S., Dergacheva E.A., et al. Global processes and the formation of global education (interdisciplinary research): collective monograph: in 2 books. / Under the general editorship of S.A. Buryanov, A.I. Krivenky. Book I. Moscow: Moscow State Pedagogical Univ., 2019. 200 p.

117. Buryanov S.A., Afanasyeva D.A. Internationally recognized right to freedom of conscience as a constitutional value in the context of global contradictions // Materials of the republican scientific and practical conference "The Constitution - as a factor in the stability of the state" (November 5, 2022). Dushanbe. 2022. 430 p. Pp. 81-85.

118. Buryanov S.A., Afanasyeva S.A., Zvonarev A.V., et al. Global processes and the formation of global education (interdisciplinary research): collective monograph: in 2 books. / Under the general editorship of S.A. Buryanov, A.I. Krivenky. Book II. Moscow: Moscow State Pedagogical Univ., 2019. 276 p.

119. Buryanov S.A., Buryanov M.S. Some problems of development of international law in the context of digital globalization // Materials of the republican scientific and practical conference "Constitution - as a factor of state stability" (November 5, 2022). Dushanbe. 2022. 430 p. Pp. 86-94.

120. Buryanov S.A., Buryanov M.S. The Future of Human Rights in the Context of the Transition to Digital Economy 4.0 // Interaction of the State and Civil Society in the Introduction of Special Measures in the Sphere of Economy: collection of materials from the International Scientific and Practical Conference / ed. T.A. Soshnikova. Moscow: Publishing House of Moscow University for the Humanities, 2022. 392 p.

121. Buryanov S.A., Buryanov M.S. Challenges of global IT technologies and some approaches to their legal regulation // Science. Entrepreneurship. Innovations: collection of materials of the International scientific and practical forum, Minsk, March 30–31, 2023 / editorial board: O. V. Bodakova [et al.]; under the general editorship of Cand. of Law, Assoc. Prof. A. N. Shklyarevsky. Minsk: BGEU, 2023. 272 p. pp. 26–30.

122. Buryanov S.A., Buryanov M.S. Global challenges of digitalization and prospects for the development of law and legal education // Russia in the 21st century: education as an important civilizational institution for the development and formation of Russian cultural and historical identity. Collection of reports andmaterials of the XXX Moiseev readings - scientific and practical conference. Moscow, 2022. 477 p. Pp. 279-285.

123. Buryanov S.A., Buryanov M.S. Global challenges of digital inequality and prospects for their legal regulation // World civilizations. 2022. Vol. 7. No. 2. P. 7-13.

124. Buryanov S.A., Buryanov M.S. Global challenges of digital technologies // Philosophy and challenges of modernity: on the 90th anniversary of the Institute of Philosophy of the National Academy of Sciences of Belarus: Proc. Int. sci. conf. (April 15-16, 2021, Minsk). In 3 vols. Vol. 2 / Institute of Philosophy of the National Academy of Sciences of Belarus; editorial board: A.A. Lazarevich (chairman) [et al.]. Minsk: Four Quarters, 2021. 378 p. Pp. 89-92.

125. Buryanov S.A., Buryanov M.S. Global challenges of digital technologies and prospects for their international legal regulation // Digital technologies and law: collection of scientific papers of the I International scientific and practical conference (Kazan, September 23, 2022) / edited by I.R. Begishev, E.A. Gromova, M.V. Zaloilo, I.A. Filippova, A.A. Shutova. In 6 volumes. Vol. 2. Kazan: Publishing house "Poznanie" of Kazan Innovation University, 2022. 556 p. Pp. 266-273.

126. Buryanov S.A., Buryanov M.S. Global crisis of information civilization and prospects for the transition to sustainable managed human-oriented development // Cultural and civilizational crisis in the context of the information society: materials of the international scientific and practical conference, Vitebsk, December 2, 2022 / Vitebsk. state University; editorial board: A.A. Lazarevich (editor-in-chief), E.V. Davlyatova, E.I. Rudkovsky. - Vitebsk: VSU named after P.M. Masherov, 2022. - 273 p. P. 217-220.

127. Buryanov S.A., Buryanov M.S. The UN Global Digital Compact as an Important Step Towards Overcoming Inequality and Achieving Social Justice // Social Justice and Law: Towards Strengthening Peace and Preventing Crises" collection of materials from the International scientific and practical conference / ed. T. A. Soshnikova. Moscow: Publishing house of Moscow University for the Humanities, 2023. 356 p. Pp. 29-33.

128. Buryanov S.A., Buryanov M.S. On the Prospects of International Legal Enshrinement of Digital Human Rights // Law. Economy. Social Partnership [Electronic resource]: Coll. of scientific papers of the International University "MITSO"; editorial board: V.F. Ermolovich (editor-in-chief) [and others]. Minsk: International University "MITSO", 2023. 1044 p. Pp. 394-397.

129. Buryanov S.A., Buryanov M.S. On the Issue of Problems and Prospects of Adopting the UN Global Digital Compact // X Legal Readings: Russian State and Law: Development Vectors Based on Traditions: All-Russian Scientific and Practical Conference (April 7–8, 2023, Syktyvkar): Collection of Articles. Part 1 / eds. V.D. Potapov, V.V. Vorobyov. Syktyvkar: Publishing House of Syktyvkar State University named after Pitirim Sorokin, 2023. 198 p. pp. 28-32.

130. Buryanov S.A., Buryanov M.S. The concept of evolutionary transition to human-oriented global governance // Century of globalization. 2021. No. 3. P. 86-100. DOI: 10.30884/vglob/2021.03.07

131. Buryanov S.A., Buryanov M.S. New threats to global security and prospects for the development of international law // Eurasian Law Journal. No. 11 (150) 2020. P. 35-40.

132. Buryanov S.A., Buryanov M.S. Prospects for human-oriented development of law in the context of modern digital global processes and challenges // Law in modern Belarusian society: collection of scientific papers: issue 16. National Center for Legislation and Legal Research of the Republic of Belarus. Minsk: Colorgrad. 2021. 832 p. P. 16-25.

133. Buryanov S.A., Buryanov M.S. Human rights in the context of digital global processes and challenges // Ontology and axiology of law. Abstracts of reports and communications of the Tenth international scientific conference. Pred. editorial board S.K. Buryakov. Omsk, 2021. 140 p. P. 87-89.

134. Buryanov S.A., Buryanov M.S. Human rights as a key factor in achieving sustainable managed development // The Age of Globalization. 2022. No. 4 (44) P.97-110. DOI: 10.30884/vglob/2022.04.07

135. Buryanov S.A., Buryanov M.S. Formation of global education as a key factor in overcoming planetary threats and the survival of human civilization // Intellectual Culture of Belarus: Problems of Interpretation of Philosophical Heritage and Modern Tasks of Humanitarian Knowledge: Proc. of the Sixth Int. sci. Conf. (November 17-18, 2022, Minsk). In 2 vols. Vol. 2 / Institute of Philosophy of the

National Academy of Sciences of Belarus; ed. board A.A. Lazarevich (chairman) [et al.]. Minsk: Four Quarters, 2022. 372 p. Pp.

136. Buryanov S.A., Buryanov M.S., Nikitaev D.M. Protection of human rights in the sphere of freedom of conscience by the mechanisms of the OrganizationThe United Nations in the context of global processes and threats // Legal World. No. 1. 2021. P. 45-49.

137. Buryanov S.A., Krivenkiy A.I. On the state and prospects of the formation of global education, including legal // State and Law. No. 8. 2019. P. 95-100. DOI 10.31857/S013207690006247-3

138. Buryanov S.A., Krivenkiy A.I. On the concept and content of globalization of education // Eurasian Law Journal. No. 8 (135). 2019. P. 15-19.

139. Buryanov S.A., Krivenkiy A.I. On the concept and content of globalization of education // Philosophy of education and modernity: on the 10th anniversary of the Department of Philosophy of Education in the structure of the Philosophy Faculty of Moscow State University. Proceedings of the scientific and practical conference with international participation. Moscow, November 22, 2018 / Ed. and compiled by E.V. Bryzgalina, V.A. Prokhoda, P.N. Moscow: Philosophical Faculty of Moscow State University, 2018. 252 p. (Electronic publication). P. 55-57.

140. Buryanov S.A., Krivenkiy A.I. On the issue of transformation of legal education in the context of globalization of law // Values and meanings. 2019. No. 5 (63). P.126–146.

141. Buryanov S.A., Krivenkiy A.I. Strategic Prospects for the Development of Law and Legal Education in the Context of Modern Global Processes // Main Trends and Prospects for the Development of Modern Law: Proceedings of the Annual International Scientific Conference in Memory of Professor Felix Mikhailovich Rudinsky, April 19, 2018 / Under the general editorship of Doctor of Law, Professor D.A. Pashentsev. Moscow: Moscow State Pedagogical Univ., Bely Veter, 2018. 506 p. Pp. 247-251.

142. Buryanov S.A., Krivenkiy A.I., Pashentsev D.A., Romanova G.V. Issues of globalization of culture and protection of cultural rights of man and citizen (interdisciplinary study): monograph. Under the general editorship of A.I. Krivenkiy, S.A. Buryanov. Moscow: Moscow State Pedagogical Univ., 2018. 200 p.

143. Buryanov S.A., Nikitaev D.M. The problem of freedom of conscience as an important component of global research // Global processes and new formats of multilateral cooperation: collection of

scientific papers of conference participants / edited by I.V. Ilyin. Moscow, MOOSIPNN N.D. Kondratieva, 2016. 282 p. P. 163-167.

144. Varlamova N.V. Digital rights - a new generation of human rights? // Proceedings of the Institute of State and Law of the Russian Academy of Sciences. 2019. No. 4. Electronic resource. URL: https://cyberleninka.ru/article/n/tsifrovye-prava-novoe-pokolenie-prav-cheloveka (date of access: 05/25/2023).

145. Getman-Pavlova I.V. International Law: textbook for universities / I.V. Getman-Pavlova, E.V. Postnikova. 3rd ed., revised and enlarged. Moscow: Yurait Publishing House, 2020. 560 p.

146. Global processes and the formation of global education (interdisciplinary research): monograph: in 2 books / auth. coll.: S.A. Buryanov, E.S. Demidenko, E.A. Dergacheva et al.; under the general editorship of S.A. Buryanov, A.I. Krivenky. Book 1. Moscow: Moscow State Pedagogical Univ., 2019. 200 p.

147. Global report Digital 2020 // Media platform Vc.ru. Electronic resource. URL: https://vc.ru/future/109699-internet-2020-v-rossii-i-mirestatistika-i-trendy (date of access: 05/25/2023).

148. Golysheva K.V., Buryanov S.A. Criminal liability for international crimes in the context of problems of achieving global security and sustainable development // Current directions of development of branches of law in the context of the new reality. Materials of the All-Russian scientific and practical conference. Edited by A.V. Semenov, T.V. Slyusarenko, V.G. Moscow, 2023. Pp. 821-826.

149. Civil Human Rights in Russia: Contemporary Theory and Practice Issues. (Edited by F.M. Rudinsky). Moscow: ZAO TF MIR, 2006. 449 p.

150. Civil human rights: modern problems of theory and practice. Volgograd. 2004. 452 p.

151. Report on the activities of the Commissioner for Human Rights in the Russian Federation for 2020. Electronic resource. URL: https://rg.ru/2021/04/01/rg-publikuet-doklad-o-deiatelnosti-upolnomochennogo-po-pravam-cheloveka-za-2020-god.html (date accessed: 05/25/2023).

152. History of aesthetics. Monuments of world aesthetic thought in 5 volumes. Vol.1, 1962. 682 p.

153. Kazimirova N.G. Formation of international legal institutions in the context of the globalization process: Diss... cad. social sciences. Moscow, 2003. 163 p.

154. Karasev A.T., Kozhevnikov O.A., Meshcheryagina V.A. Digitalization of legal relations and its impact on the implementation of individual constitutional rights of citizens in the Russian Federation // Antinomies.2019. No. 3. Electronic resource. URL: https://cyberleninka.ru/article/n/tsifrovizatsiya-pravootnosheniy-i-ee-vliyanie-na-realizatsiyu-otdelnyh-konstitutsionnyh-prav-grazhdan-v-rossiyskoy-federatsii (date of access: 05/25/2023).

155. Kartkhia A.A. Civil-law model of regulation of digital technologies: Diss... Doc. of Law. M. 2019. 394 p.

156. Katsura A.V., Mazur I.I., Chumakov A.N. Planetary humanity: on the edge of the abyss. M. 2016. 208 p.

157. Kolosov Yu.M. International law: textbook / Yu.M. Kolosov, E.S. Krivchikova. Moscow: Higher education, Yurait-Iedat. 2009. 1012 p.

158. Kolosov, Yu. M., Krivchikova E. S. International Law: textbook / ed. A. N. Vylegzhanin. Moscow: Higher education, Yurait-Iedat. 2009. 1012 p.

159. Constitutional rights and freedoms of man and citizen in the Russian Federation. Ed.: Tiunov O.I. M.: Norma, 2005. 591 p.

160. Kuksin I.N., Khoda, V.D. Digitalization - a new reality in law and new threats // Theory of State and Law. 2020. No. 4 (20). P. 115-128.

161. Kutsobina E.V. Globalization as a general scientific problem: Diss... Cand. Philosophical Sciences. Moscow, 2005. 193 p.

162. Lenshin S.I. The Impact of Digitalization on the Economic and Legal Regime of Strengthening the Defense and Security Capability of Russia // Step into the Future: Artificial Intelligence and Digital Economy. Revolution in Management: New Digital Economy or New World of Machines. Proceedings of the II International Scientific Forum. General editor P.V. Terelyansky. M. 2018. Pp. 85-90.

163. Lenshin S.I. Improving the legal regime of digitalization of the Russian economy to ensure its defense and security // Step into the future: artificial intelligence and digital economy: Smart Nations: Economy of digital equality. Proceedings of the III International Scientific Forum. 2020. P. 337-341.

164. Lee S.M. Environmental component of the concept of sustainable development: International legal aspects: Diss... Cand. of Law. M. 2004. 182 p.

165. Lukashuk I.I. Interaction of international and domestic law in the context of globalization // Journal of Russian Law. Moscow: Norma, 2002, No. 3. P. 115-128.

166. Lukashuk I.I. International Law. General Part: Textbook for Students of Law Faculties and Universities. - 3rd Edition, Revised and Supplemented. Moscow: Wolters Kluwer, 2005. 432 p.

167. Marchenko M.N. State and Law in the Context of Globalization. / M.N. Marchenko. M. 2009. 400 p.

168. International Covenants on Human Ashes: Value Characteristics / Proceedings of the International Scientific and Practical Conference: Ed. T.A. Soshnikova, N.V. Kolotova – M. Publishing House of Moscow University for the Humanities, 2016. 312 p.

169. Morozova L.A. Theory of State and Law: textbook / L.A. Morozova. 4th ed., revised and enlarged. Moscow: Eksmo, 2010. 510 p.

170. Mutagirov D. Z. Human rights and freedoms: textbook for universities / D. Z. Mutagirov. 2nd ed., corrected and supplemented. Moscow: Yurait Publishing House, 2020. 516 p.

171. Normative regulation of the digital environment. URL: https://www.economy.gov.ru/material/directions/gosudarstvennoe_upravlenie/normativnoe_regulirovanie_cifrovoy_sredy/ (date of access: 25.05.2023).

172. Pavlenko E.M. Human rights education as a basis for the formation of legal culture and human rights culture in the Russian Federation / E.M. Pavlenko. Moscow: 2016. 216 p.

173. Palazyan A.S. Legal behavior of an individual in the conditions of formation of civil society in modern Russia: Diss... Cand. Philosophical Sciences. Rostov-on-Don, 2000. 133 p.

174. Surveillance Pandemic. How the Authorities Spy on Russians in the Age of Coronavirus - International Agora Report. Electronic resource. URL: https://spy.runet.report/ (date accessed: 25.05.2023).

175. Pashentsev D.A. Main trends in the influence of modern digital technologies on the development of law // Law and education. 2019. No. 7. P. 4-9.

176. Podzigun I.M. Globalization and global problems: philosophical and methodological analysis: Diss... Doctor of Philosophy. sciences. M. 2003. 384 p.

177. Political Dictionary // Encyclopedias and Dictionaries. Electronic resource. URL: http://enc-dic.com/ (date accessed: 05/25/2023).

178. Polozhikhina M.A. Information and digital inequality as a new type of socio-economic differentiation of society // Economic and social problems of Russia / M.A. Polozhikhina. 2017. No. 2. P. 119-142.

179. Human and civil rights and freedoms: theoretical aspects and legal practice: materials of the annual International scientific conference in memory of Professor Felix Mikhailovich Rudinsky, April 27, 2017 / edited by D.A. Pashentsev. Ryazan: Concept Publishing House, 2017. 348 p.

180. Human Rights / Ed. Corresponding Member of the Russian Academy of Sciences E.A. Lukasheva. Moscow: Norma, 2004. 573 p.

181. Human Rights in a Changing World / Proceedings of the International Scientific and Practical Conference: eds. T.A. Soshnikova, E.A. Karpov, N.V. Kolotova. Moscow: Moscow University for the Humanities, Human Rights in a Changing World. 2017. 376 p.

182. Human Rights. United Nations Decade for Human Rights Education (1995-2004) No. 4. ABC of Teaching Human Rights. Practical Activities in Primary and Secondary Schools. Geneva: b.i., 2003. 186 p.

183. Law in the digital world. URL: https://rg.ru/2018/05/29/zorkin-zadacha-gosudarstva-priznavat-i-zashchishchat-cifrovye-prava-grazhdan.html (date of access: 25.05.2023).

184. Rassolov I.M., Chubukova S.G., Mikurova I.V. Biometrics in the context of personal data and genetic information: legal issues // Lex Russica. 2019. No. 1 (146). URL: https://cyberleninka.ru/article/n/biometriya-v-kontekste-personalnyh-dannyh-i-geneticheskoy-informatsii-pravovye-problemy (accessed: 25.2023).

185. Dictionary of environmental terms and definitions // Encyclopedias and dictionaries. Electronic resource. URL: http://enc-dic.com/ (date accessed: 05/25/2023).

186. Strategic Psychology of Globalization: Psychology of Human Capital. Glossary / edited by A. I. Yuryev. Moscow, 2006. 511 p.

187. Talapina E.V. Evolution of human rights in the digital age // Proceedings of the Institute of State and Law of the Russian Academy of Sciences. 2019. No. 3. Electronic resource. URL: https://cyberleninka.ru/article/n/evolyutsiya-prav-cheloveka-v-tsifrovuyu-epohu (date of access: 05/25/2023).

188. Tarasova A.G. Legal procedures and implementation of human rights: theoretical and legal aspect: Diss... candidate of legal sciences. M. 2012.206 p.

189. Theory of State and Law: Textbook for Universities. edited by Marchenko M.N. Moscow: "Zertsalo". 2004. 800 p.

190. Tolchinsky M.V. Risks of globalization: Diss... Cand. Phil. Sciences. Moscow, 2012. 155 p.

191. Ursul A. D. Global governance: evolutionary perspectives // Century of globalization. 2014. No. 1. P. 16-28.

192. Ursul A.D. Globalization of Law and Global Law: Conceptual and Methodological Problems // Law and Politics. 2012. No. 8. P. 1284-1297.

193. Ursul A.D. Education for Sustainable Development: First Results, Problems and Prospects // Sociodynamics. 2015. No. 1. P. 11-74.

194. Ursul A.D. Advanced Education. From Modernization to Futurization. Saarbrücken: Dictus Publishing, 2015. 304 p.

195. Farkhutdinov I.Z. Iranian Doctrine of Preventive Self-Defense and International Law // Eurasian Law Journal. 2017. No. 1. P. 15-26.

196. Farkhutdinov I.Z. Iranian Doctrine of Preventive Self-Defense and International Law (end) // Eurasian Law Journal. 2017. No. 2. P. 15-25.

197. Farkhutdinov I.Z. International or global law // International lawyer / I.Z. Farkhutdinov, 2004. No. 4. P. 15-23.

198. Farkhutdinov I.Z. International or global law // International lawyer. All-Russian journal of international law. 2004. No. 4. P. 15-23.

199. Federal Ombudsman: There is a lack of legal instruments to protect citizens' digital rights. Electronic resource. URL: https://ombudsmanrf.org/news/novosti_upolnomochennogo/view/federalnyj_ombudsmen:_dlja_zashhity_cifrovykh_prav_grazhdan_sushhestvuet_nedostatok_pravovykh_instrumentov (date of access: 05/25/2023).

200. Philosophical Dictionary. Electronic resource. URL: http://slovariki.org/filosofskij-slovar/14089 (date of access: 05/25/2023).

201. Foucault M. Intellectuals and Power. Selected political articles, speeches and interviews. Part 1. Trans. from French.S. Ch. Ofertas, general editor. / M. Foucault; V. P. Vizgin and B. M. Skuratov. –M.: Praxis, 2002. 384 p.

202. Chumakov A.N. Globalization. Contours of the Integral World: monograph. 3rd ed., revised and enlarged. Moscow: Prospect, 2017. 448 p.
203. Chumakov A.N. Global world: the problem of management // Century of globalization. 2010. No. 1. P. 3-15.
204. Chumakov A.N. Global World: Clash of Interests Monograph / A.N. Chumakov. M. 2018. 512 p.
205. Chumakov A.N. Metaphysics of globalization. Cultural and civilizational context. Monograph. 2nd ed., corrected and enlarged. Moscow: Prospect, 2017. 496 p.
206. Chumakov A.N. World politics in the context of globalization // Asia and Africa today. 2016. No. 12 (713). P. 71-74.
207. Chumakov A.N. On the nature of instability in the modern world // Credo New. 2016. No. 3 (87). P. 8.
208. Chumakov A.N. Main trends of global development: realities and prospects // Century of globalization. 2018. No. 4(28). P. 3-15. DOI: https://doi.org/10.30884/vglob/2018.04.01
209. Chumakov A.N. The problem of management as a reason for discussion // The Age of Globalization. 2012. No. 2. P. 35-42.
210. Chumakov A.N. Path to Philosophy. Works of Different Years. M. 2021. 608 p.
211. Chumakov A.N. Theory and practice of solving global problems. M. 2015. 352 p.
212. Chumakov A.N., Stark L.P. The Club of Rome: on the results of half a century of activity // The Age of Globalization. 2019. No. 4 (32). P. 40-49.
213. Ethics and "digital": ethical problems of digital technologies. Analytical report. 2020. Electronic resource. URL: http://ethics.cdto.center/ (date of access: 05/25/2023).
214. Legal Dictionary // Encyclopedias and Dictionaries URL: http://enc-dic.com/ (date accessed: 05/25/2023).
215. IV International Scientific Congress "Globalistics-2015": Prospects for the Development of International Law in the Context of Globalization of Public Relations // Eurasian Law Journal.No. 10 (89). 2015. pp. 348-353.
216. Bradeley JS (ed.). Killing by Remote Control: The Ethics of an Unmanned Military. NY: Oxford University Press, 2013.
217. BurianovM. Here's why we need a Declaration of Global Digital Human Rights. World Economic Forum. Electronic resource. URL:

https://www.weforum.org/agenda/2020/08/here-s-why-we-need-a-declaration-of-global-digital-human-rights/ (accessed: 25.05.2023).

218. Burlamaqui J.-J. The Principles of Natural and Political Law [1747]. 2010. P. 19

219. Chamayou G. Théorie du drone.Paris: La Fabrique. 2013.

220. Davidson JD, Rees-Mogg W. The Sovereign Individual: Mastering the Transition to the Information Age. Simon & Schuster. 1999. – 448 p.

221. Harari, YN Homo Deus: A Brief History of Tomorrow. Harvill Secker. 2016. 448 p.

222. Floridi, L. Philosophy and Computing: An introduction. Routledge. 1999. – 257 p.

223. European System for the Protection of Human Rights / ed. by R. Macdonald, P. Matscher and H. Petzold. Dordrecht ; Boston; London, 1993. P. XXII.

224. Global Digital Human Rights for 4IR.Electronic resource. URL: http://maxlaw.tilda.ws/digitalhumanrights (date of access: 05/25/2023).

225. Global Law Forum: an online community focused on achieving sustainable development. Electronicresource. URL: http://maxlaw.tilda.ws/ (date of access: 05/25/2023).

226. Governance of Artificial Intelligence risk.Electronic resource. URL: https://globalchallenges.org/global-risks/artificial-intelligence/governance-of-artificial-intelligence-risk/ (date of access: 05/25/2023).

227. Measuring trends in Artificial Intelligence.Electronic resource. URL: https://aiindex.stanford.edu/report/ (date of access: 05/25/2023).

228. OECD. Axer le secteur public sur les données: marche à suivre.2020 196 p. ISBN: 9789264673489

229. Scharre P. Army of None: Autonomous Weapons and the Future of War. NY: WW Norton&Company, 2018.

230. Schwab K. Shaping the Fourth Industrial Revolution.Portfolio Penguin. 2018. 288 rub.

231. Sitdikova, RI, Sitdikov.RB Digital rights as a new type of property rights [Tsifrovyye vid imushchestvennykh prav]. Property relations in the Russian Federation, 9 (in Russian). 2018.

232. SlaughterA.M. A New World Order. Princeton and Oxford: Princeton University Press, 2004. 341 p.

233. Serres, M. La petite poucette. Éditions Le Pommier. 2012. – 80 p.

233. Digital human rights in the context of global processes: theory and practice of implementation URL: https://lib.dm-centre.ru/lib/document/gpntb/ESVODT/32e4fb24a3c5ea20a777647f8b94be80/ (date accessed: 05/25/2023).

234. The International Dimensions of Human Rights / K. Vasak, General Editor. Vol. 1. Paris, 1982.R. XVI

235. Lessig, L. Code and Other Laws of Cyberspace. NewYork. 1999. – 424 p.

APPLICATIONS

APPLICATION 1. Project of the Master of Law, alumnus of the youth group of the CIS countries Generation Connect (GC-CIS) of the UN ITU, member of the international community Global Shapers of the World Economic Forum, coordinator of the scientific and educational project and international initiative Global Digital Human Rights for the 4IR, Youth Ambassador and Envoy of the SDGs of Russia 2020-2021, expert of the Youth Council under the Commissioner for Human Rights in the Russian Federation Burianov Maksim Sergeevich

The goal of the project is to implement and protect human rights in the digital space and to increase the level of digital literacy among citizens. We create a legal condition for responding to the digital threats of our time (digital inequality, AI risks, total surveillance and violation of human rights), ensuring the use of technologies for good. We are pioneers in the field of rethinking human rights in the context of modern digital transformation (at the same time, the project cooperates with the Youth Council under the Commissioner for Human Rights of the Russian Federation, the Youth Group of the CIS Countries (GC-CIS) of the

UN ITU, with the World Economic Forum Global Shapers Moscow, the UN Information Center as part of the activities on the UN SDG agenda, where we present our expert positions and implement legal, scientific, project and educational initiatives).

Context. Rapid changes in people's lives (social relations) require constant updating of the law (law) that regulates it. This means that the digitalization of all spheres of life and the associated digital risks (inequality, militarization, total control, etc.) require adequate legal regulation. Since in the modern world law is inextricably linked with human rights, digital human rights are a response to the dark sides of digitalization. This requires updating the international legal obligations of states in the field of human rights in the digital age.

The current principles in the field of human rights were developed and enshrined in fundamental documents from 1948 to 2000 (the Universal Declaration of Human Rights and others). At that time, scientific and technological progress was in the transition stage from the second industrial revolution (electricity, mass production) to the third (automation and computerization).

The survey showed that 97.7% of respondents (Figure 1) from various regions of the world (Australia, Argentina, Brunei, Brazil, Great Britain, Germany, Greece, Cameroon, Mexico, Mali, Russia, Switzerland, the USA, etc.) believe that the adoption and implementation of the Declaration of Global Digital Human Rights can create conditions for a human-centered direction of 4IR development and overcoming global threats (digital inequality, digital wars, digital dictatorships, etc.).

Do you consider it appropriate to develop, adopt and implement the Declaration of Global Digital Human Rights as a tool to address global digital threats?

130 ответов

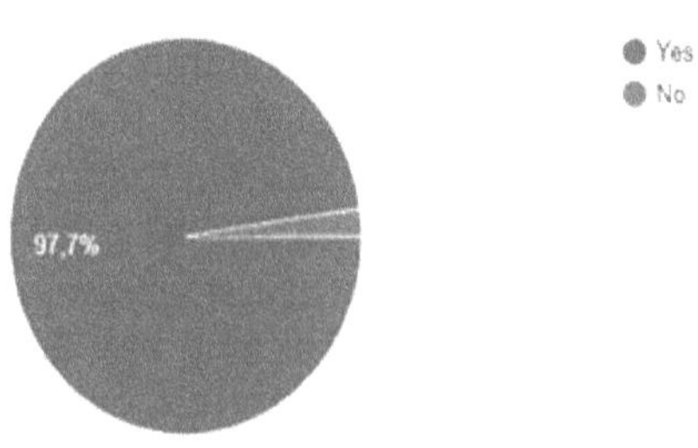

Figure 1 - Survey by Global Shapers Moscow (an initiative of the World Economic Forum) on the need to consolidate and implement digital human rights http://maxlaw.tilda.ws/digitalhumanrights

As part of the scientific stage, we processed a large block of statistical, factual and scientific information from Russian and foreign sources on human rights, the Fourth Industrial Revolution, globalization 4.0 and digitalization risks. In 2020, together with the Global Law Forum and Global Shapers Moscow WEF, we conducted a survey in Russia on the need for digital human rights. The answers are shown in Figure 2, as we can see, more than 80% of respondents confirmed the need to consolidate and subsequently implement digital human rights to overcome digital threats 4.0 at the federal and global levels.

Figure 2 - Global Shapers Moscow survey (an initiative of the World Economic Forum) on the need to consolidate and implement digital human rights in Russia http://maxlaw.tilda.ws/digitalhumanrights

Overall, the survey showed the importance of initiating a Declaration of Global Digital Human Rights to overcome global digital challenges, and subsequently, on this basis, a petition was launched in support of the document.

To implement the concept, a project was launched jointly by WEF Global Shapers Moscow and the Global Law Forum. The concept was tested at scientific conferences (IMEMO, MSU, MGIMO, IZISP, RUDN and others), which are reflected on the Global Law Forum website http://maxlaw.tilda.ws/globalnews and became the basis of the program of the Ambassador of the Sustainable Development Goals of Russia - Maksim Burianov.

Results and achievements achieved within the framework of the implementation and/or functioning of the described activity, project, product:

Step 1. Scientific study of global processes and threats in the context of technological progress (2017-2020). The result was the scientific testing of the concept of global law and global digital human rights. The following threats were identified: global digital inequality, digital militarization, digital surveillance, AI threats.

Step 2. A methodology for assessing the impact of digitalization on human life and human rights has been developed (May - July 2020), the project is entering a practice-oriented stage. As part of the project, a team of shapers developed a toolkit for assessing the impact of digitalization (4IR) on human life. The survey showed that 97% of respondents (about 200 people) consider it necessary to adopt the Declaration as a response to global digital risks (digital militarization, digital inequality, digital surveillance and AI threats). July 2020.

Step 3. The project became the basis for the activities of the UN Sustainable Development Goals Ambassador (May - July 2020). Digital human rights began to be considered as a guarantee of the implementation of the UN SDG agenda in Russia and the world.

Step 4. The project was supported within the framework of the World Economic Forum agenda and Maxim Buryanov wrote a draft document of the Declaration of Global Digital Human Rights. You can read the article at the link: https://www.weforum.org/agenda/2020/08/here-s-why-we-need-a-declaration-of-global-digital-human-rights/.

Step 5. A petition in support of the draft Declaration of Global Digital Human Rights was launched (September - December 2020). The petition in support of the Declaration was signed by more than 60 youth organizations from all over the world.

Step 6. The project is presented at the international and all-Russian level in competitions and conferences, winning many prizes (2019-2021). The project has won many scientific competitions in Eurasia and the CIS, articles on the concept have been published in the journals YADRO-RINTS and VAK (5 pieces). The ideas of the project formed the basis of the Davos Lab agenda from Global Shapers (digital access).

Step 7. Project on the UN ITU youth agenda (April - October 2021). The project is presented within the framework of the youth expert group at the UN International Telecommunication Union (at the regional and global level on youth dimension), the development of international human rights is included in the document on recommendations for the UN ITU from the youth working group.

Step 8. The project is presented at the international BRIF forum (September-October 2021).

The project is presented at the V International Baikal Risk Forum BRIF-2021, more details at the link: https://www.youtube.com/watch?v=qGduvby-Aqk.

The concept of digital human rights has gained recognition within the financial community (which will lead to the development of ESG 4.0 – non-financial reporting for technology companies).

Step 9. Educational events were held at Moscow State University (April-October 2021).

A number of seminars on sustainable development and digital human rights were held for the Faculty of Global Processes at Moscow State University, for more details, please follow the link: https://www.youtube.com/watch?v=gg-gkdwOX24&t=2478s.

Step 10. The monograph "Digital Human Rights in the Context of Modern Global Processes: Prospects for International Legal Enshrinement" was published and presented at various venues (February-April 2022).

Step 11. Digital human rights in focus of the Youth Council under the Commissioner for Human Rights of the Russian Federation. A National Strategy for the implementation of digital rights was proposed. At a meeting of the Youth Council under the Commissioner for Human Rights in the Russian Federation, Maxim Buryanov presented the concept of the National Strategy for the implementation of digital human rights in the Russian Federation.

Step 12: My proposals on youth rights were included in the 2022 Generation Connect Youth Call to Action My Digital Future manifesto.

Step 13. The project was approved - I defended my master's thesis on the topic "Digital Rights of the Child" in the context of generative AI, web 3.0, neural interfaces and metaverses. The book Digital Rights of the Child was published.

Step 14. A modern digital state should be human-centric. Legal regulation of the digital environment related to human rights will ensure that technologies serve the benefit of citizens, increase citizens' trust in digital service systems and minimize abuses (for example, excessive data collection and human rights violations). It was also proposed to develop a sustainable digital transformation of a human-oriented state, which will accelerate the process of involving personnel for the digital transformation of states. It was also proposed to consider the digital state as a factor in solving global problems and the need to achieve the UN Sustainable Development Goals.

APPLICATION 2. Declaration digital rights of children.

Introduction

Digital technologies have become integral to every facet of life, profoundly shaping the experiences and opportunities available to children worldwide. While these advancements offer unprecedented opportunities for learning, creativity, and connection, they also introduce new risks that can impact the well-being and development of young individuals. Recognizing both the transformative potential and the challenges posed by the digital age, we, the representatives of the international community, hereby establish the **Declaration of Global Digital Rights of the Child**. This Convention aims to safeguard and promote the digital rights of children, ensuring their protection, development, and active participation in the evolving digital landscape.

Preamble

Acknowledging the pervasive influence of digital technologies on the lives of children and understanding the critical need to balance innovation with protection, we affirm the following:

- **Digital Technologies as Integral Tools**: Digital technologies are fundamental in modern childhood, offering avenues for education, self-expression, and global engagement.
- **Protection and Development**: It is imperative to protect children from digital risks while fostering their growth and development in a secure and supportive digital environment.
- **Global Cooperation**: Effective realization of digital rights requires coordinated efforts across states, organizations, and communities to establish and uphold universal standards.
- **Human-Centric Approach**: The design and implementation of digital technologies must prioritize the rights, autonomy, and well-being of children, ensuring that technology serves as a tool for empowerment rather than exploitation.

Basic Principles

1. **Equal Access to Digital Rights**
    - All children, irrespective of age, socio-economic status, or geographical location, have the right to equal opportunities to access and utilize digital technologies,

bridging the digital divide and fostering inclusive participation.

2. **Protection from Digital Surveillance**
   - Children are entitled to the protection of their personal data, including biometric, medical, and genetic information. Digital surveillance must be transparent, accountable, and must not infringe upon the rights and freedoms of children.
3. **Protection from Digital Militarization**
   - Children have the right to be safeguarded from the use of digital technologies for military purposes. This includes preventing the deployment of artificial intelligence and other technologies in ways that could harm or exploit children.
4. **Bridging the Digital Divide**
   - Ensuring that all children have access to the necessary digital resources to meet their developmental needs and aspirations, thereby promoting equitable growth and opportunities.
5. **Use of Techno-Legal Platforms**
   - Children have the right to engage with digital products that adhere to principles of transparency, accountability, and robust data protection, ensuring ethical and legal use of technologies.
6. **Principle of Transparency and Information**
   - Children have the right to be informed about how their personal data is collected, used, and protected. Any changes to data usage policies must be communicated clearly and accessibly.
7. **Data Minimization Principle**
   - The collection and utilization of data pertaining to children should be limited to what is strictly necessary, ensuring the minimization of data exposure and enhancing privacy protections.
8. **Principle of Security and Data Protection**
   - Children have the right to secure personal data that is protected from unauthorized access, misuse, and breaches, ensuring their digital interactions are safe and private.
9. **Principle of Consent and Control**

- Children, along with their parents or guardians, should have the authority to control the use of their personal data, including providing informed consent for data collection and processing.

Articles

Article 1. Digital Rights of Children

1. **Comprehensive Digital Rights**: Every child possesses digital rights that include equal access to digital technologies, freedom of expression in digital spaces, protection of personal data and privacy, digital education and literacy, and protection from harmful influences and exploitation online.
2. **Legal Enshrinement**: These rights are legally enshrined to ensure the survival, development, and active participation of children in the digital world. They recognize the unique ways in which children interact with technology and aim to mitigate the uncertainties and risks associated with digital advancements.
3. **Key Digital Rights**:
    1. **Right to Privacy and Data Protection**
    2. **Right to Equal Access and Participation in the Global Internet**
    3. **Right to Education and Development**
    4. **Right to Security and Protection**
    5. **Right to Equality and Inclusion**
    6. **Right to Ethical and Legal Use and Development of Technologies**
    7. **Right to Protection from Digital Inequality and Discrimination**
    8. **Right to Transparency and Control Over the Use of Data** (including personal, genetic, and biometric data)
    9. **Right to Ethical and Legal Use and Development of Artificial Intelligence**
    10. **Right to Know Digital Rights and Freedoms in an Accessible Form**
    11. **Right to Informed Consent and Data Protection** in contexts such as neural interfaces, neurotechnologies, neuromarketing, and big data systems
    12. **Right to Be Forgotten**: The right to have non-essential data erased upon reaching adulthood, ensuring that

only critical medical data is retained in decentralized registries.

Article 2. Equal Access and Opportunities

States and societies are obligated to ensure that all children, regardless of their social, economic, or geographical circumstances, have equal access to digital technologies. This includes providing the necessary infrastructure, resources, and support to bridge the digital divide and promote inclusive digital participation.

Article 3. Freedom of Expression and Participation

Children have the right to freely express their thoughts, opinions, and ideas in digital environments, within the bounds of social, cultural, and ethical norms. Additionally, children should be empowered to actively participate in digital spaces, influencing digital policies and contributing content that shapes the digital future.

Article 4. Protection of Personal Data and Confidentiality

Children's personal data must be safeguarded in all digital interactions. Data collection, processing, and usage must adhere to stringent data protection laws, ensuring that children and their guardians have control over personal information and are informed about data handling practices.

Article 5. Education and Digital Literacy

Children have the right to comprehensive digital education that equips them with the skills and knowledge to navigate the digital world safely and ethically. This includes training in digital technologies, digital safety, critical thinking, and ethical use of digital resources.

Article 6. Protection from Harmful Influences and Exploitation

1. **Safeguarding Well-being**: Children must be protected from harmful digital content and exploitation, including online abuse, harassment, discrimination, and illegal activities.
2. **Zero Digital Challenges**: All stakeholders must strive to eliminate digital challenges that threaten children's rights and well-being.
3. **Hierarchical Digital Risks**:
    1. **Complex Cyber Risks**: Addressing cyber threats related to the Internet and smartphones in the context of the third industrial revolution.
    2. **Universal Rights Risks**: Mitigating risks to children's universal rights in the fourth industrial revolution.
    3. **Digital Divide**: Bridging gaps in access and participation.

4. **Digital Surveillance**: Preventing intrusive and unauthorized surveillance.
5. **Digital Militarization**: Protecting against the militarization of digital technologies.
6. **Syncretic Digital Risks**: Combating blended and multifaceted digital threats.
7. **Cognitive Digital Risks**: Addressing impacts on cognitive development and mental health.
8. **Existential Risks of Technologies**: Mitigating overarching threats posed by advanced technologies.

Article 7. Right to Access Technology and the Global Internet
Children are entitled to access modern digital technologies and the global Internet, ensuring they can participate fully in the digital age. This right encompasses access to Industry 4.0 technologies, tailored to different age groups for optimal platform adaptation and usage.

Article 8. Right to Co-Creation with Artificial Intelligence
Children have the right to engage with and co-create alongside artificial intelligence, fostering their development and enabling them to harness the potential of generative AI technologies. This collaboration supports metacognitive development and prepares children for active participation in a technologically advanced society.

Article 9. Right to Digital Sovereignty
Upon reaching adulthood, children should have access to decentralized digital technologies that empower their digital sovereignty. This includes participation in decentralized governance models and access to technologies that support autonomous and non-violent societal structures.

Article 10. International Cooperation and Coordination
States, international organizations, technology companies, and civil society must collaborate to realize children's digital rights globally. International bodies should establish mechanisms and standards to monitor the implementation of digital rights, facilitate the exchange of best practices, and support the development of a cohesive global framework for digital human rights.

Conclusion

We call upon all states, societies, and stakeholders to adopt and effectively implement this **Declaration of Global Digital Rights of the Child**. By doing so, we ensure the protection and development of children's digital rights in the digital age, fostering a safe, inclusive, and ethical digital environment. This environment will empower children to realize their full potential, actively participate in shaping the digital future, and contribute to a sustainable and equitable global society.

Key Theses on Global Digital Human Rights

1. **Expansion of Human Rights in the Digital Era**: Global digital human rights encompass traditional rights and new rights such as digital autonomy, data protection, technology access, and prevention of algorithmic discrimination.
2. **Human Rights by Design**: Integrating human rights principles into the architectural and coding phases of AI development ensures compliance at the engineering level, preventing violations by default.
3. **Global Digital Commons**: Creating open-access digital resources fosters equal access to knowledge and technology, supporting inclusive development.
4. **Polycentric Governance Model**: A multi-level governance approach involves states, corporations, NGOs, academic communities, and local groups to ensure transparency and accountability in digital rights enforcement.
5. **Interdisciplinary Legal Architecture**: Developing international treaties, flexible regulatory sandboxes, and supranational standards to adapt legal norms to rapidly evolving technologies.
6. **Audit and Control Institutions**: Establishing the Global Digital Rights Authority (GDRA) for independent auditing, certification, and monitoring of AI systems.
7. **Economic Incentives and ESG 4.0**: Implementing ESG 4.0 metrics and economic incentives like tax breaks and grants to encourage companies to uphold high standards of digital human rights.
8. **Technological Standards and Cryptographic Protocols**: Developing open standards, APIs, and libraries to ensure human rights compliance in AI, using blockchain audits and Zero-Knowledge Proofs for transparency and accountability.
9. **Inclusivity and Cultural Adaptation**: Adapting digital rights to diverse cultural contexts through syncretic digital culture councils and local protocols, ensuring respect for local values and traditions.

10. **Education and Digital Literacy**: Incorporating courses on digital rights, AI ethics, and neuroethics into educational programs, creating global educational initiatives and meta-universities to enhance awareness and competencies.
11. **Prevention of Digital Militarization**: Establishing international technological verification institutions and strategic consultations to prevent AI misuse in military applications and human rights violations.
12. **Feedback and Adaptation Mechanisms**: Regularly revising norms through foresight studies and multi-stakeholder committees, ensuring legal frameworks remain relevant and adaptable to technological advancements.
13. **Support for Vulnerable Groups**: Ensuring access and protection for children, low-income populations, people with disabilities, and minority groups through specialized tools and programs.
14. **Decentralized Control Mechanisms**: Implementing DAO guilds and decentralized courts for cyber disputes, ensuring neutral and transparent decisions regarding digital rights violations.
15. **Cognitive Well-being and Psychohygiene**: Developing cognitive well-being metrics and integrating AI tools to support users' emotional health, preventing manipulation and cyberbullying.
16. **Integration of Neuroethics and Bioethics**: Embedding ethical principles into the development of neurointerfaces and biotechnologies to respect human autonomy and prevent dehumanization.
17. **Open Hardware Solutions**: Encouraging the development of open hardware platforms for AI with verified security to reduce monopolization risks and ensure transparency in technological processes.
18. **International Digital Treaties**: Adopting global digital treaties, including digital human rights in the UN agenda, forming the foundation for sustainable digital development and cooperation.
19. **Ethical Engineering and Sustainable Innovations**: Incorporating ethical engineering principles in technology development, supporting sustainable innovations aimed at enhancing quality of life and protecting human rights.

20. **Technological Ecolife**: Promoting love and care for the digital ecosystem, striving for harmonious coexistence between humans and technology, ensuring ecological and social sustainability of digital systems.

Pathways for Comprehensive Integration of Concepts
To effectively embed digital rights into AI systems and broader societal structures, the following comprehensive integration pathways are recommended:

1. **Developing and Implementing Libraries and Modules**
   - **human_rights_guard**: Create an open-source Rust library integrated into all AI projects, containing a set of formally verified policy modules to enforce digital rights.
   - **digital_rights_integration**: Develop a Python library facilitating interaction with the Rust module and external audit and policy update systems.
2. **Creating and Supporting Institutional Structures**
   - **Establishing MADCOR**: Form the International Agency for Digital Rights (MADCOR) to independently audit AI systems, issue compliance certificates, and monitor digital rights adherence.
   - **Organizing Regular Forums and Consultations**: Conduct regular forums and consultations involving all stakeholders to update standards and policies collaboratively.
3. **Implementing Economic and Regulatory Incentives**
   - **Tax Incentives and Grants**: Introduce tax benefits and grants for companies that adhere to digital human rights standards.
   - **Integrating ESG 4.0**: Embed ESG 4.0 metrics into corporate reporting and investment ratings to ensure accountability and encourage compliance.
4. **Educational Initiatives and Cultural Adaptation**
   - **Incorporating Digital Rights into Education**: Integrate digital rights, AI ethics, and neuroethics into educational curricula at all levels.
   - **Forming Syncretic Digital Culture Councils**: Create councils to adapt global digital rights to various cultural

contexts, ensuring respect for local values and traditions.

5. **Developing Technological Standards and Protocols**
   - **Open Standards Creation**: Develop and disseminate open standards, APIs, and libraries to ensure AI systems comply with human rights principles.
   - **Blockchain Audit Integration**: Utilize blockchain for transparent and accountable auditing of AI systems.
6. **Monitoring and Adaptation Mechanisms**
   - **Regular Policy Reviews**: Conduct policy reviews every 6-12 months through foresight studies and stakeholder committees to ensure legal frameworks remain current and adaptable.
   - **Feedback Systems**: Implement crowdsourced feedback platforms and DAO guilds for continuous improvement of digital rights frameworks.
7. **Supporting and Protecting Vulnerable Groups**
   - **Developing Specialized Tools**: Create tools and programs to ensure digital technology access and protection for children, low-income populations, individuals with disabilities, and minority groups.
   - **Integrating Digital Psychogiene**: Develop AI tools to support users' emotional well-being and prevent cyberbullying and manipulation.
8. **Aligning with Global Initiatives and Collaboration**
   - **Incorporating into UN SDGs**: Embed digital human rights into the United Nations Sustainable Development Goals (SDGs) to ensure global alignment and support.
   - **Support for Global Digital Commons**: Promote open-access digital resources to ensure equitable knowledge and technology distribution.

APPLICATION 3. Digital Compass: A Guiding Framework for Human-Centric Technology Use

1. Introduction

The **Digital Compass** serves as your navigator in the intricate digital landscape. Designed to guide individuals of all ages, it provides comprehensive recommendations for understanding and utilizing digital technologies responsibly. By integrating essential aspects such as digital rights, cognitive impacts, usage rules, and ethical interactions, the Digital Compass ensures that technology serves humanity's best interests. This framework aligns with the principles of human rights by design and facilitates a seamless transition from centralized societal structures to non-hierarchical, networked, and non-violent frameworks at local, national, regional, and global levels.

2. Principles

The foundation of the Digital Compass is built upon six core principles that ensure technology serves humanity ethically and responsibly:

1. **Digital Human and Child Rights**
   - **Universal Access and Protection**: Every individual, regardless of age, has the right to access, use, create, and distribute information and technology. They are also entitled to protection from misuse, ensuring their digital interactions uphold their fundamental rights.
2. **Neurohygiene and Critical Thinking**
   - **Cognitive Awareness**: Understanding how digital technologies influence our cognitive processes is crucial. Encouraging critical thinking helps individuals evaluate information effectively, promoting healthier and more informed technology use.
3. **Adaptability and Personalization**
   - **User-Centric Design**: Technologies must adapt to the unique needs and preferences of each user, ensuring personalized experiences that enhance rather than hinder personal and professional growth.
4. **Technology for Good**
   - **Ethical Utilization**: Technology should be leveraged to improve the quality of life, address societal challenges, and foster positive outcomes for individuals and communities.
5. **Minimization of Personal Data**

- **Privacy by Design**: Strive to minimize the collection and storage of personal data, enhancing user privacy and security while reducing the risk of data breaches and misuse.

6. **Security and Cybersecurity**
   - **Protective Measures**: Educate users on the fundamentals of cybersecurity to safeguard themselves and their data, ensuring secure interactions in the digital realm.

3. Digital Compass for Technology

This section presents a curated list of relevant technologies tailored to specific age groups. Each technology is examined for its potential uses, cognitive impacts, usage guidelines, and alignment with digital human rights.

3.1 Age-Based Technology Recommendations

Users can select their age group to access relevant technologies and corresponding guidelines:

- **A) Ages 5 to 7**
- **B) Ages 7 to 14**
- **C) Ages 14 to 18**
- **D) Ages 18 to 26**
- **E) Ages 26 to 60**
- **F) Ages 60 and above**

Each subsection provides detailed insights into specific technologies applicable to the selected age group.

3.1.1 Technology: Artificial Intelligence Chatbot (e.g., ChatGPT)

Description and Potential Use: Artificial Intelligence chatbots like **ChatGPT**, developed by OpenAI, utilize machine learning to generate coherent and contextually relevant text. These technologies have diverse applications across various fields:

- **Education**: Assisting students with homework, providing explanations, and facilitating interactive learning experiences.
- **Healthcare**: Offering preliminary medical advice, mental health support, and patient education.
- **Business**: Enhancing customer service through automated responses, improving efficiency in handling inquiries.
- **Entertainment**: Creating interactive storytelling experiences, engaging users in conversational games.

Cognitive Impact and Risks: AI chatbots can significantly influence cognitive processes:

- **Benefits**:
  - **Enhanced Learning**: Provide immediate information and support, aiding in knowledge acquisition.
  - **Skill Development**: Foster critical thinking and problem-solving skills through interactive dialogues.
- **Risks**:
  - **Illusion of Understanding**: Users might overestimate the chatbot's comprehension, leading to misinformation.
  - **Reduced Human Interaction**: Overreliance on AI for communication may diminish interpersonal skills.
  - **Data Privacy**: Potential misuse of personal data if not adequately protected.

Usage Rules and Principles: To ensure safe and ethical use of AI chatbots:

1. **Digital Hygiene**: Protect personal information by avoiding sharing sensitive data.
2. **Critical Evaluation**: Assess the accuracy and reliability of information provided by the chatbot.
3. **Time Management**: Limit the duration of interactions to prevent overuse and dependency.
4. **Ethical Engagement**: Use chatbots for constructive purposes, avoiding requests that promote harm or unethical behavior.

Implementation of Digital Human Rights: AI chatbots can uphold several digital human rights:

- **Right to Access Information**: Provide timely and accurate information to users.
- **Right to Privacy**: Ensure data protection through robust security measures.
- **Right to Equality and Inclusion**: Offer unbiased and inclusive interactions, preventing discrimination.
- **Right to Informed Consent**: Ensure users are aware of data usage policies and obtain consent before data collection.

---

4. List of All Current Digital Human Rights

The Digital Compass enumerates a comprehensive list of digital human rights, encompassing both traditional and emerging rights essential for the digital age. These rights include:

1. **Right to Privacy and Data Protection**

2. **Right to Equal Access to and Participation in the Global Internet**
3. **Right to Education and Development**
4. **Right to Security and Protection**
5. **Right to Equality and Inclusion**
6. **Right to Legal Use and Development of Technologies**
7. **Right to Protection from Digital Inequality and Discrimination**
8. **Right to Transparency and Control Over the Use of Data** (personal, genetic, and biometric data)
9. **Right to Legal Use and Development of Artificial Intelligence**
10. **Right to Know Your Digital Rights and Freedoms in an Accessible Form**
11. **Right to Informed Consent and Data Protection** in contexts such as neural interfaces, neurotechnologies, neuromarketing, and big data systems.
12. **Right to Be Forgotten** for non-essential data retained only in decentralized registries until adulthood.
13. **Right to Quality Global Education in the Digital Age** in co-creation with AI and metacognitive development.
14. **Right to Startup Development**
15. **Right to Digital Security and Protection from Existential Threats**
16. **Right to Digital Sovereignty and Participation in Governance**
17. **Right to Digital Consciousness and Digital Autonomy**
18. **Right to Digital Responsibility and Ethics**
19. **Right to Unconditional Basic Income**
20. **Right to Free Social Services**
21. **Right to Open Source Technologies**

5. List of All Relevant Parties Influencing the Digital World

Effective implementation and protection of digital human rights require the involvement of various stakeholders, including:

- **Public Sector**: Governments, international organizations, regulatory bodies.
- **Private Sector**: Corporations, startups, venture funds, commercial partners.
- **Education Sector**: Universities, schools, academic communities.
- **Media Sector**: Traditional media outlets, social networks, new media platforms.

- **Global Community**: Children, youth, parents, caregivers, adults, vulnerable groups.
- **Non-Governmental Organizations (NGOs)**: Advocacy groups, digital rights organizations.
- **Technologists and Developers**: AI researchers, software engineers, data scientists.
- **Legal and Ethical Experts**: Lawyers, ethicists, policy makers.

6. Digital Literacy and Skills

Digital literacy encompasses the ability to effectively and ethically navigate the digital world. Key components include:

- **Basic Digital Skills**: Understanding how to use digital devices, software, and the internet.
- **Advanced Technical Skills**: Proficiency in programming, data analysis, and AI interaction.
- **Critical Thinking**: Evaluating the credibility of digital information and recognizing biases.
- **Digital Hygiene**: Practices to protect personal data and maintain cybersecurity.
- **Ethical Usage**: Awareness of the ethical implications of technology use and responsible behavior online.

7. Digital Ethics and Digital Citizenship

Digital ethics and citizenship focus on fostering responsible and respectful behavior in the digital environment. Core elements include:

- **Respect for Privacy**: Valuing and protecting personal and others' privacy.
- **Integrity**: Maintaining honesty and transparency in digital interactions.
- **Respect for Intellectual Property**: Acknowledging and adhering to copyright and licensing laws.
- **Inclusivity**: Promoting diversity and preventing discrimination in digital spaces.
- **Accountability**: Taking responsibility for one's actions and their impacts in the digital realm.
- **Civic Engagement**: Participating actively and ethically in digital governance and community-building.

8. Resources, Tools, and Applications

A variety of resources, tools, and applications are available to support the principles and practices outlined in the Digital Compass:

- **Educational Platforms**: Online courses, tutorials, and workshops on digital literacy and ethics.

- **Security Tools**: Antivirus software, VPNs, password managers, and encryption tools.
- **Privacy Applications**: Data anonymization tools, privacy-focused browsers, and secure communication apps.
- **Ethical AI Frameworks**: Guidelines and toolkits for developing and implementing ethical AI systems.
- **Digital Rights Advocacy**: Platforms and organizations dedicated to promoting and protecting digital human rights.
- **Generative Knowledge Systems**: AI-driven platforms that support continuous learning and co-creation with AI.

Conclusion

The **Digital Compass** is an essential tool for navigating the digital landscape responsibly and ethically. By adhering to its principles and guidelines, individuals and organizations can ensure that digital technologies enhance human well-being, uphold digital human rights, and foster a sustainable and inclusive digital ecosystem. Embedding digital rights into code and integrating them into societal structures facilitates a human-centric digital transformation, promoting equity, security, and active participation in the evolving digital world.

APPLICATION 4. General Comment No. 25 (2021) on children's rights in relation to the digital environment (Committee on the Rights of the Child prepared for the Convention on the Rights of the Child)

I. Introduction

1. Children consulted for this general comment reported that digital technologies are vital to their lives now and for their future: "with digital technologies we can access information from all over the world"; "[digital technologies] have introduced me to core aspects of my identity"; "when you are sad, the Internet can help you see something that brings you joy"[143].

2. The digital environment is constantly evolving and expanding, encompassing information and communication technologies, including digital networks, content, services and applications, connected devices and environments, virtual and augmented reality, artificial intelligence, robotics, automated systems, algorithms and data analytics, biometrics and implant technology.[144].

3. The digital environment is becoming increasingly important in most aspects of children's lives, including in times of crisis, as public functions such as education, public services and commerce increasingly rely on digital technologies. It offers new opportunities to realise children's rights, but also creates the risk of their rights being violated or undermined. During the consultations, children expressed the view that the digital environment should support, encourage and protect their safe and equal participation: "we would like the government, technology companies and teachers to help us control misinformation online"; "I would like "I would like to get a clear picture of what is actually happening with my data... Why is it being collected? How is it being collected?"; "I am... concerned about the dissemination of my data"[145].

---

[143] Summary report on the consultations with children carried out during the preparation of this general comment "Our rights in a digital world", pp. 14 and 22. URL:https://5rightsfoundation.com/uploads/Our%20Rights%20in%20a%20Digital%20World.pdfAll references to children's views refer to this report.

[144] A glossary of terms can be found on the Committee's website:https://tbinternet.ohchr.org/_layouts/15/treatybodyexternal/Download.aspx?symbolno=INT%2fCRC%2fINF%2f9314&Lang=en.

[145] "Our rights in a digital world", pp. 14, 16, 22 and 25.

4. The rights of every child in the digital environment must be respected, protected and fulfilled. Digital innovations have a wide-ranging and interdependent impact on children's lives and rights, even when children themselves do not have access to the Internet. Meaningful access to digital technologies can help children realize the full range of their civil, political, cultural, economic and social rights. However, if digital technologies are not universally accessible, existing inequalities are likely to be reinforced, and new ones may emerge.

5. The present general comment is based on the Committee's experience in examining States parties' reports and holding a day of general discussion on digital media and children's rights, the jurisprudence of the human rights treaty bodies, the recommendations of the Human Rights Council and the special procedures of the Council, two rounds of consultations with States, experts and other stakeholders on a concept paper and preliminary draft, and international consultations with 709 children living in a variety of settings in 28 countries in several regions of the world.

6. This general comment should be read in conjunction with the Committee's other relevant general comments and its guidelines on the implementation of the Optional Protocol to the Convention on the sale of children, child prostitution and child pornography.

II. Target

7. In the present general comment, the Committee explains how States parties should implement the Convention in the digital environment and provides guidance on appropriate legislative, policy and other measures to ensure full compliance with their obligations under the Convention and its Optional Protocols, in the light of the opportunities, risks and challenges posed by the digital environment promotion, respect, protection and fulfilment of all children's rights in the digital environment.

III. General principles

8. The following four principles serve as a lens through which the implementation of all other rights under the Convention should be viewed. They should guide the determination of the measures needed to ensure the realization of children's rights in the digital environment.

A. Non-discrimination

9. The right to non-discrimination requires States Parties to ensure that all children have equal, effective and meaningful

access to the digital environment[146]. States parties should take all necessary measures to overcome digital exclusion. This includes ensuring free and safe access for children in designated public spaces and investing in policies and programs that promote affordable access to and wise use of digital technologies for all children in educational settings, communities and at home.

10. Children may be discriminated against because they are excluded from using digital technologies and services, or because they receive messages containing hate speech, or because they are treated unfairly when using these technologies. Other forms of discrimination may arise when automated processes that result in filtering, profiling or decision-making are based on biased, partial or unlawfully obtained data about a child.

11. The Committee calls upon States parties to take proactive measures to prevent discrimination on the basis of gender, disability, socioeconomic status, ethnic or national origin, language or any other grounds, as well as discrimination against minority and indigenous children, asylum-seeking, refugee and migrant children, lesbian, gay, bisexual, transgender and intersex children, child victims of trafficking or sexual exploitation, children in alternative care, children deprived of their liberty, andother groups of children in vulnerable situations. Specific measures will be needed to close the gender-related digital divide for girls and to ensure that special attention is paid to issues such as access, digital literacy, privacy and online safety.

B. The best interests of the child

12. The best interest of the child is an evolving concept that requires context-specific assessment.[147]. The digital environment was not originally designed for children, but it plays a significant role in their lives. States parties should ensure that in all activities related to the creation, regulation, design, use and management of the digital environment, the best interests of every child are a primary consideration.

13. States parties should involve national and local bodies responsible for monitoring children's rights in such activities. When considering the best interests of the child, they should take into account all the rights of children, including their right to seek, receive and

---

146 General Comment No. 9 (2006), paras. 37–38.

147 General Comment No. 14 (2013), para. 1.

impart information, to be protected from harm and to have their views duly taken into account, and ensure transparency in the assessment of the best interests of the child and the criteria applied.

C. The right to life, survival and development

14. The opportunities provided by the digital environment play an increasingly important role in the development of children and can be critical to their lives and survival, particularly in crisis situations. States parties should take all necessary measures to protect children from risks of violation of their right to life, survival and development. Risks related to content, communications, conduct and contracts relate to, inter alia, violent and sexual content, cyberbullying and harassment, gambling, exploitation and abuse, including of a sexual nature, and the promotion or incitement of suicide or life-threatening activity, including by criminals or armed groups listed as terrorist and extremist organizations. States should identify and address new risks to children arising in a variety of contexts, including by listening to their views on the nature of the specific risks they face.

15. The use of digital devices should not cause harm, nor should it replace face-to-face interactions between children or between children and parents or caregivers. States parties should pay particular attention to the impact of technology during the early years of a child's life, when brain plasticity is at its greatest and the social environment, in particular relationships with parents and caregivers, is critical for their cognitive, emotional and social development. Precautions may be needed during the early years, depending on the design, purpose and uses of technologies. Parents, caregivers, educators and other relevant actors should be trained and given guidance on the appropriate use of digital devices, considering research on the impact of digital technologies on child development, particularly during critical periods of neurodevelopment in early childhood and adolescence.[148].

D. Respect for the child's opinion

16. Children reported that the digital environment provides them with critical opportunities to have their views heard on issues that affect them.[149]. The use of digital technologies can help ensure

---

148 General Comment No. 24 (2019), para. 22; and General Comment No. 20 (2016), pp. 9–11.

149 "Our rights in a digital world", p. 17.

children's participation at local, national and international levels[150]. States parties should promote awareness of and access to digital means for children to express their views, and provide training and support for children to participate on an equal basis with adults, anonymously where necessary, so that they can effectively claim their rights both individually and as a group.

17. In developing legislation, policies, programmes, services and training on children's rights in relation todigital environment, States parties should involve all children, listen to their needs and give due consideration to their views. They should ensure that digital service providers actively engage with children, provide appropriate safeguards, and take children's views into account when developing products and services.

18. States parties are encouraged to use the digital environment to consult children on relevant legislative, administrative and other measures and to ensure that their views are given serious weight and that children's participation does not result in undue monitoring or data collection that violates their right to privacy, freedom of thought and opinion. They should ensure that consultation processes include children who do not have access to or skills in using technology.

IV. Developing abilities

19. States parties should respect the evolving capacity of the child as a principle of law-making, which envisages a process through which children progressively acquire knowledge, understanding and capacity to act[151]. This process is particularly important in the digital environment, where children may participate more independently, subject to supervision by parents and caregivers. The risks and opportunities associated with children's participation in the digital environment vary according to their age and stage of development. States parties should be guided by these considerations whenever they develop measures to protect children in this environment or to facilitate their access to it. Age-appropriate measures should draw on the best and most current research from a range of disciplines.

20. States parties should take into account the changing situation of children and their role in the modern world, the uneven experiences and knowledge of children in different areas of skills and

---

150 General Comment No. 14 (2013), paras. 89–91.

151 General Comment No. 7 (2005), para. 17; and General Comment No. 20 (2016), paras. 18 and 20.

activities, and the varying nature of the risks involved. These considerations should be balanced against the importance of realizing their rights in a supported environment and the diversity individual experience and circumstances[152]. States parties should ensure that digital service providers offer services that are appropriate to the evolving capacities of children.

21. In accordance with the duty of States to provide appropriate assistance to parents and caregivers in the performance of their child-rearing responsibilities, States parties should promote awareness among parents and caregivers of the need to respect children's growing autonomy, their evolving capacities and their privacy. They should support parents and caregivers in acquiring digital literacy and understanding the risks to children so that they can help children to realize their rights in the digital environment, including the right to protection.

V. General measures of implementation to be taken by States Parties

22. To ensure that children's rights can be exercised and protected in the digital environment, a wide range of legislative, administrative and other measures, including precautionary measures, are required.

A. Legislation

23. States parties should review, adopt and update national legislation in line with international human rights standards to ensure that the digital environment is compatible with the rights enshrined in the Convention and its Optional Protocols. Legislation should remain relevant in the context of technological advances and new practices. They should provide for the mandatory use of child rights impact assessments to take into account children's rights in legislation, budget allocations and other administrative decisions related to the digital environment, and promote their use by government agencies and businesses involved in the digital environment[153].

B. Integrated Policy and Strategy

24. States parties should ensure that national strategies on children's rights specifically address the digital environment and implement relevant regulations, industry codes, design standards and

---

[152] General Comment No. 20 (2016), para. 20.

[153] General Comment No. 5 (2003), para. 45; General Comment No. 14 (2013), para. 99; and General Comment No. 16 (2013), paras. 78–81.

action plans, which should be regularly assessed and updated. Such national strategies should aim to enable children to benefit from the digital environment and ensure their safe access to it.

25. Child online protection should be integrated into national child protection policies. States parties should implement measures to protect children from risks, including cyberbullying and digitally and online child sexual exploitation and abuse, ensure the investigation of such crimes and provide remedies and support to child victims. They should also address the needs of children in disadvantaged or vulnerable situations, including by providing child-friendly information, where appropriate, translated into relevant minority languages.

26. States parties should ensure that effective child online protection mechanisms and appropriate strategies to ensure their safety, while respecting other children's rights, are in place wherever children have access to the digital environment, including in the home, educational institutions, internet cafés, youth centres, libraries, and health and alternative care settings.

C. Coordination

27. To address the comprehensive implications of the digital environment for children's rights, States parties should designate a government body to coordinate policies, guidelines and programs relating to children's rights between central and local authorities.[154]. Such a national coordination mechanism should engage with schools and the ICT sector and collaborate with business, civil society, academia and organizations to realize children's rights in relation to the digital environment at cross-sectoral, national, regional and local levels.[155]It should draw, as appropriate, on technological and other relevant expertise from government and other bodies and be independently assessed for the effective performance of its duties.

D. Resource allocation

28. States parties should mobilize, allocate and utilize public resources to implement legislation, policies and programmes aimed at the full realization of children's rights in the digital environment and increasing universal accessibility to the digital environment, which is necessary to address the growing impact of the digital environment on

---

[154] General Comment No. 5 (2003), para. 37.

[155] Ibid., pp. 27 and 39.

children's lives and to promote equal access to and affordability of Internet services and connectivity.[156].

29. Where resources are provided by the business sector or through international cooperation, States parties should ensure that the exercise of their own powers, the processes of mobilizing their revenues, the allocation and expenditure of budgetary funds are not subject to interference or disruption as a result of the actions of third parties[157].

E. Data collection and research

30. Regularly updated data and research are critical to understanding the implications of the digital environment for children's lives, analysing its impact on their rights and assessing the effectiveness of public measures. States parties should undertake adequately resourced collection of reliable, comprehensive data disaggregated by age, sex, disability, geographic location, ethnic and national origin and socioeconomic status. Such data and research, including research with and by children, should inform legislation, policies and practices and be publicly available.[158]When collecting data and conducting research on children's digital lives, their privacy and the highest ethical standards must be respected.

F. Independent monitoring

31. States parties should ensure that the mandates of national human rights institutions and other relevant independent institutions cover children's rights in the digital environment and that they are able to receive, investigate and take appropriate action on complaints from children and their representatives.[159]. Where independent oversight bodies exist to monitor activities related to the digital environment, national human rights institutions should work closely with such bodies to ensure that they effectively fulfil their mandate with regard to children's rights.[160].

G. Dissemination of information, awareness raising and training

32. States parties should disseminate information and conduct awareness-raising campaigns on children's rights in the digital

---

156 General Comment No. 19 (2016), para. 21.

157 Ibid., subparagraph 27 b).

158 General Comment No. 5 (2003), paras. 48 and 50.

159 General Comment No. 2 (2002), paras. 2 and 7.

160 Ibid., p. 7.

environment, with particular attention to those actors whose actions have a direct or indirect impact on children. States parties should promote educational programmes for children, parents and caregivers, the general public and policymakers to enhance their knowledge of children's rights in relation to the opportunities and risks associated with digital products and services. Such programmes should include information on how children can benefit from digital products and services and develop their digital literacy and skills, how to protect children's privacy and prevent victimization, and how to recognize and take appropriate action when a child is being harmed online or offline. Such programmes should be based on research and consultation with children, parents and caregivers.

33. Professionals working with and for children, as well as in the business sector, including the technology industry, should receive training that includes an examination of how the digital environment impacts on children's rights inhow children exercise their rights in the digital environment and how they access and use technologies in different contexts. They should also be trained in the application of international human rights standards in the digital environment. States parties should ensure that initial and in-service training on the digital environment is provided to professionals at all levels of the education system in order to develop their knowledge, skills and practices.

H. Cooperation with civil society

34. States parties should systematically involve civil society, including child-led groups and non-governmental organizations working on children's rights, as well as organizations working on digital issues, in the development, implementation, monitoring and evaluation of laws, policies, plans and programmes relating to children's rights. They should also ensure that civil society organizations can carry out their activities related to the promotion and protection of children's rights in the digital environment.

I. Children's Rights and the Business Sector

35. The business sector, including non-profit organizations, directly and indirectly impacts children's rights when providing services and products related to the digital environment. Businesses must respect children's rights and prevent and remedy violations of their rights in connection with the digital environment.

Member States are obliged to ensure that business enterprises meet these obligations.[161].

36. States parties should take measures, including through the development, monitoring, implementation and evaluation of legislation, regulations and policies, to ensure that businesses comply with their obligations to ensure that their networks or online services are not used in a way that results in or contributes to the violation or abuse of children's rights, including their right to privacy and protection, and to provide prompt and effective remedies to children, parents and caregivers. They should also encourage businesses to make publicly available information and accessible and timely recommendations to support safe and beneficial digital activities for children.

37. States parties have an obligation to protect children from business-related violations of their rights, including the right to protection from all forms of violence in the digital environment. While business-related activities may not be directly involved in harmful conduct, their activities may cause or contribute to violations of children's right to freedom from violence, including through the development and provision of digital services. States parties should adopt, monitor and enforce laws and regulations to prevent violations of the right to protection from violence, as well as laws and regulations to investigate, adjudicate and remedy violations committed in connection with the digital environment.[162].

38. States parties should require the business sector to carry out due diligence with regard to children's rights, in particular by conducting child rights impact assessments and making their findings fully public, with particular attention to the differentiated and sometimes severe impact of the digital environment on children.[163]. They should take appropriate measures to prevent, monitor, investigate and punish violations of children's rights by business enterprises.

39. In addition to developing laws and policies, Member States should require all businesses that impact children's rights in the digital environment to comply with regulations, industry codes and terms of service that meet the highest standards of ethics, privacy and safety in the design, engineering, development, provision, distribution

---

161 General Comment No. 16 (2013), paras. 28, 42 and 82.

162 Ibid., p. 60.

163 Ibid., pp. 50 and 62–65.

and marketing of their products and services. This includes businesses that directly target children, have children among their end users or otherwise impact children. They should require such businesses to adhere to high standards of transparency and accountability and encourage them to take measures to ensure that the best interests of the child are taken into account when innovating. In addition, they should require such businesses to provide age-appropriate instructions to children or to parents or caregivers of young children regarding the terms and conditions of their services.

J. Commercial Advertising and Marketing

40. The digital environment includes businesses that rely financially on the processing of personal data to target revenue-generating or paid content, and such processes intentionally and unintentionally impact children's digital experiences. Many of these processes involve multiple commercial partners, a chain of commercial activities, and the processing of personal data that may result in the abuse or infringement of children's rights, including through the promotion of design elements that predict a child's actions and direct them toward more extreme content, automated notifications that may interrupt sleep, or the use of a child's personal information or location information to direct potentially harmful commercially oriented content.

41. When regulating advertising and marketing directed to and accessible to children, Member States should give primary consideration to the best interests of the child. Sponsorship, product placement and all other forms of commercial content should not perpetuate gender or racial stereotypes and there should be a clear distinction between them and all other content.

42. Member States should prohibit by law the profiling or targeting of children of any age for commercial purposes based on a digital record of their actual or inferred characteristics, including group or collective data, targeting through profiling based on association or similarity of interests. Exposing children, directly or indirectly, to practices based on neuromarketing, emotion analysis, immersive advertising, and advertising in virtual and augmented reality to promote products, apps and services should also be prohibited.

K. Access to justice and legal remedies

43. Children face particular challenges in accessing justice in the digital environment across a range of issues reasons. These difficulties arise, in particular, from the lack of legislation establishing sanctions for violations of children's rights specifically in connection

with the digital environment, the difficulty of obtaining evidence or identifying violators, or the fact that children and their parents or caregivers are unaware of their rights or what constitutes a violation or infringement of their rights in the digital environment. Further problems may arise if children are required to disclose information about sensitive or confidential activities online or if they fear retaliation from peers or social exclusion.

44. States parties should ensure that appropriate and effective judicial and non-judicial mechanisms for redress for violations of children's rights in the digital environment are widely disseminated and easily accessible to all children and their representatives. Complaint and reporting mechanisms should be free, secure, confidential, responsive, child-friendly and in accessible formats. States parties should also provide for the possibility of filing collective complaints, including class actions and public interest litigation, and providing legal or other appropriate assistance, including through specialized services, to children whose rights have been violated in or through the digital environment.

45. States parties should establish referral mechanisms and provide effective support to child victims of such violations, coordinate their activities and regularly monitor and evaluate them.[164]. These mechanisms should include measures for the identification, treatment and aftercare of child victims and their social reintegration. Training in the identification of child victims should be part of the referral mechanism, including for digital service providers. The measures provided for in such mechanisms should be multi-agency and child-friendly, in order to prevent the repeated and secondary victimization of the child in the context of investigations and judicial proceedings. In this regard, special measures may be required to protect confidential information and to provide redress for harm related to the digital environment.

46. Appropriate reparations include restitution, compensation and satisfaction and may include an apology, remedial action, removal of illegal content, access to psychological rehabilitation services or other measures.[165]. With regard to violations in the digital environment, remedies should take into account the vulnerability of

---

164 General comment No. 21 (2017), para. 22. See also General Assembly resolution 60/147, annex.

165 General Comment No. 5 (2003), para. 24.

children and the need to promptly stop current harm and prevent future harm. States parties should ensure that violations do not recur, including by reforming relevant laws and policies and effectively implementing them.

47. Digital technologies further complicate the investigation and prosecution of crimes against children, which may have a cross-border dimension. States parties should consider how the use of digital technologies may facilitate or impede the investigation and prosecution of crimes against children and take all available preventive, law enforcement and judicial remedies, including in cooperation with international partners. They should provide specialized training to law enforcement officials, prosecutors and judges on child rights violations specifically related to the digital environment, including through international cooperation.

48. Children may face particular difficulties in obtaining remedies for violations of their rights in the digital environment by businesses, particularly in the context of their global operations.[166] States parties should consider taking measures to ensure that children's rights are respected, protected and fulfilled in the context of extraterritorial business activities and operations, provided that there is a reasonable connection between the State and the conduct in question. They should ensure that businesses establish effective complaint mechanisms; however, such mechanisms should not hinder children's access to State remedies. They should also ensure that bodies with oversight powers over children's rights, including those dealing with health and safety, data protection and consumer rights, education, and advertising and marketing, investigate complaints and provide appropriate remedies where children's rights are violated or abused in the digital environment.[167]

49. States parties should provide children with child-sensitive and age-appropriate information, using child-friendly language, about their rights, as well as about reporting and complaint mechanisms, services and remedies available to them in the event of violations or abuses of their rights in the digital environment. Such information should also be provided to parents, caregivers and professionals working with and for children.

---

166 General Comment No. 16 (2013), paras. 66–67.

167 Ibid., pp. 30 and 43.

VI. Civil rights and liberties

A. Access to information

50. The digital environment provides children with a unique opportunity to exercise their right to access information. In this regard, mass media and communications, including digital and online content, play an important role.[168]. States parties should ensure that children have access to information in the digital environment and that restrictions on the exercise of this right are limited to those cases where this is provided by law and necessary for the purposes set out in article 13 of the Convention.

51. States parties should ensure and support the creation of digital content for children that is age-appropriate and empowering in accordance with their evolving capacities, and guarantee their access to a variety of information, including information on culture, sports, arts, health, civil and political issues and children's rights, held by public authorities.

52. States parties should promote the production and dissemination of such content using a variety of formats and from a diversity of national and international sources, including news media, broadcasters, museums, libraries, and educational, scientific and cultural organizations. They should, in particular, strive to increase the provision of diverse, accessible and useful content for children with disabilities and children belonging to ethnic, linguistic, indigenous and other minorities. Access to relevant information in languages that children can understand can have a significant positive impact on achieving equality.[169].

53. States parties should ensure that all children are aware of the availability of diverse and quality information on the Internet, including content that is not influenced by commercial or political interests, and can easily find it. They should ensure that automated information searching and filtering, including recommendation systems, do not prioritize paid content with commercial or political motivations over content chosen by children or at the expense of children's right to information.

54. The digital environment may include content containing gender stereotypes, discriminatory and racist language,

---

[168] General Comment No. 7 (2005), para. 35; and General Comment No. 20 (2016), para. 47.

[169] General Comment No. 17 (2013), para. 46; and General Comment No. 20 (2016), paras. 47–48.

violence, exploitation and pornography, as well as falsehoods, disinformation and misinformation, and information that encourages children to engage in illegal or harmful activities. Such information may come from a variety of sources, including other users, commercial content creators, sex offenders or armed groups listed as terrorist or extremist organizations. States parties should protect children from harmful and problematic content and ensure that relevant businesses and other digital content providers develop and implement guidelines that guarantee children's safe access to a variety of content, recognizing their right to information and freedom of expression, while protecting them from such harmful material in accordance with their rights and developing abilities[170]. Any restrictions on the operation of any Internet-based, electronic or other information dissemination system must comply with the provisions of Article 13 of the Convention.[171]. States Parties shall not intentionally interfere with, or permit others to interfere with, the supply of electricity, cellular telephone networks or the Internet in any geographic area, either partially or totally, in a manner that is likely to impair a child's access to information and communications.

55. States parties should encourage providers of digital services used by children to use concise and understandable content labels, such as age-appropriate or trustworthy. They should also encourage them to provide accessible guidance, training, educational materials and feedback/reporting mechanisms for children, parents and caregivers, educators and relevant professional groups.[172]Age- or content-based recommendation systems designed to protect children from age-inappropriate content must be implemented in a manner that respects the principle of data minimization.

56. Member States should ensure that digital service providers comply with relevant guidelines, standards and codes[173]and compliance with rules on lawful, necessary and proportionate content moderation. Content control, school filtering systems and other safety-oriented technologies should not be used to restrict children's access to information in the digital environment; they should only be used to

---

170 General Comment No. 16 (2013), para. 58; and General Comment No. 7 (2005), para. 35.

171 Human Rights Committee, general comment No. 34 (2011), para. 43.

172 General Comment No. 16 (2013), paras. 19 and 59.

173 Ibid., pp. 58 and 61.

prevent children from accessing harmful materials. Content moderation and control should be balanced with the right to protection from violations of other children's rights, in particular their right to freedom of expression and privacy.

57. Professional codes of conduct established by news media and other relevant organizations should include guidelines on how to report on digital risks and opportunities for children. Such guidelines should lead to fact-based reporting, without disclosing the identity of child victims, and in line with international human rights standards.

B. Freedom of expression

58. Children's right to freedom of expression includes the right to seek, receive and impart information and ideas of all kinds through the media of their choice. Children reported[174]that the digital environment offers great opportunities to express their ideas, opinions and political views. For children in disadvantaged or vulnerable situations, interacting with others who share their experiences using modern technology can help them express themselves.

59. Any restrictions on children's right to freedom of expression in the digital environment, such as filters, including security measures, must be lawful, necessary and proportionate. The rationale for such restrictions must be transparent and communicated to children in age-appropriate language. States parties should provide children with information and training on how to effectively exercise this right, in particular on how to create digital content and how to disseminate it safely, while respecting the rights and dignity of others and without violating laws, such as those relating to incitement to hatred and violence.

60. When children express their political or other views and express their identity in the digital environment, they may be subject to criticism, hostility, threats or punishment. States parties should protect children from cyberbullying and threats, censorship, data leaks and surveillance using digital technologies. Children should not be held accountable for expressing their opinions in the digital environment, unless they violate restrictions provided for by criminal law and consistent with article 13 of the Convention.

---

174 "Our rights in a digital world", p. 16.

61. Given the existence of commercial and political motives to promote particular worldviews, States parties should ensure that the use of automated information filtering, profiling, marketing and decision-making processes does not replace, limit or hinder children's ability to form and express their opinions in the digital environment.

C. Freedom of thought, conscience, belief and religion

62. States parties should respect the right of the child to freedom of thought, conscience and religion in the digital environment. The Committee recommends that States parties adopt or update data protection regulations and design standards to identify, define and prohibit practices that manipulate children's right to freedom of thought and opinion in the digital environment or interfere with it, for example through emotion or opinion analysis. Automated systems may make inferences about a child's internal state. States parties should ensure that automated or information filtering systems are not used to negatively influence or affect children's behaviour or emotions or to limit their capabilities or development.

63. States parties should ensure that children are not penalized because of their religion or belief or that their future opportunities are not otherwise restricted. Children's right to manifest their religion or belief in the digital environment may be subject only to legitimate, necessary and proportionate restrictions.

D. Freedom of association and peaceful assembly

64. The digital environment can provide children with the opportunity to develop their social, religious, cultural, ethnic, sexual and political identities and to participate in their communities and public life for the purposes of debate, cultural exchange, social cohesion and diversity[175]Children reported that the digital environment provides them with valuable opportunities to meet, exchange and discuss opinions with peers, decision makers and others who share their interests.[176].

65. States parties should ensure, through laws, regulations and policies, that the right of children to participate in organizations that operate partly or exclusively in the digital environment is protected. There should be no restrictions on children

---

175 General Comment No. 17 (2013), para. 21; and General Comment No. 20 (2016), paras. 44–45.

176 "Our rights in a digital world", p. 20.

exercising their right to freedom of association. and peaceful assemblies in the digital environment, except for lawful, necessary and proportionate ones[177]. Such participation should not in itself lead to negative consequences for these children, such as exclusion from school, limitation or deprivation of future opportunities, or police registration. Such participation should be safe, confidential, and free from surveillance by public or private organizations.

66. Public presence and networking opportunities in the digital environment can also support children's activism and empowerment as human rights defenders. The Committee recognizes that the digital environment enables children, including child human rights defenders, as well as children in vulnerable situations, to communicate with each other, advocate for their rights and form associations. States parties should support them, including by facilitating the creation of dedicated digital spaces, and ensure their safety.

E. Right to privacy

67. Privacy is vital for the freedom of action, dignity and safety of children, and for the exercise of their rights. Children's personal data is processed for the purpose of providing them with educational, medical and other services. Threats to children's privacy may arise from the collection and processing of data by government agencies, businesses and other organizations, as well as from criminal activities such as identity theft. Threats may also come from children's own activities, as well as from the activities of their family members, peers or others, such as when parents post photos online or an unknown person posts information about the child.

68. The data may include information about, among other things, children's identity, activities, location, communications, emotions, health and relationships. Certain combinations of personal data, including biometric data, may uniquely identify a child. Digital techniques such as automated data processing, profiling, behavioural targeting, mandatory identity verification, information filtering and mass surveillance are becoming commonplace. Such practices may result in arbitrary or unlawful interference with children's right to privacy; they may have adverse effects on children that may continue to affect them later in life.

---

177 Human Rights Committee, general comment No. 37 (2020), paras. 6 and 34.

69. Interference with a child's privacy is permissible only if it is neither arbitrary nor unlawful. Therefore, any such interference must be provided by law, pursue a legitimate aim, comply with the principle of data minimization, be proportionate and aimed at ensuring the best interests of the child and must not be contrary to the provisions, aims or objectives of the Convention.

70. States parties should take legislative, administrative and other measures to ensure that children's privacy is respected and protected by all entities and in all environments that process their data. Legislation should provide for strong safeguards, transparency, independent oversight and access to remedies. States parties should require that digital products and services that affect children have a built-in privacy feature. They should review privacy and data protection legislation on a regular basis and ensure that procedures and practices help prevent intentional or accidental violations of children's privacy. Where encryption is considered an appropriate means, States parties should consider taking appropriate measures toto identify and report cases of child sexual exploitation and abuse or child sexual abuse material. Such measures should be strictly limited in accordance with the principles of legality, necessity and proportionality.

71. In cases where consent to data processing is requested

child, States parties should ensure that consent is informed and freely given by the child or, depending on the age and evolving capacities of the child, by the child's parent or caregiver, and is obtained before those data are processed. Where the child's own consent is considered insufficient and parental consent is required for the processing of the child's personal data, States parties should require entities processing such data to verify whether such consent is informed, meaningful and given by the child's parent or caregiver.

72. States parties should ensure that children and their parents or caregivers can easily access stored data, correct inaccurate or outdated data and delete data unlawfully or unnecessarily stored by public authorities, private persons or other bodies, subject to reasonable and lawful restrictions.[178]. They should also ensure that children have the right to withdraw their consent and to object to the processing of personal data where the data controller cannot demonstrate compelling

---

178 Human Rights Committee, general comment No. 16 (1988), para. 10.

legitimate grounds for such processing. They should also provide children, parents and caregivers with information on such matters in child-friendly language and in accessible formats.

73. Children's personal data should be accessible only to authorities, organizations and persons authorized by law to process them, subject to procedural safeguards such as regular reviews and accountability measures.[179]. Data about children collected for specific purposes in any institution, including offender records in digital format, should be protected and collected solely for those purposes and should not be unlawfully or unjustifiably retained or used for any other purpose. Where information is provided in one institution and could legitimately benefit a child through its use in another institution, such as in the context of schools and higher education, the use of such data should be transparent, accountable and subject to the consent of the child, parent or carer, as appropriate.

74. Privacy and data protection laws and measures should not arbitrarily restrict children's other rights, such as their right to freedom of expression or protection. Member States should ensure that data protection legislation respects children's privacy and protects personal data in the digital environment. With continuous technological innovation, the digital environment is expanding to include more services and products, such as clothing and toys. As the places where children spend their time become "connected" through the use of embedded sensors linked to automated systems, Member States should ensure that products and services that facilitate such environments are subject to robust data protection measures and other privacy-related norms and standards. These include public spaces such as streets, schools, libraries, sports and entertainment venues and offices, including shops and cinemas, as well as the home.

75. Any surveillance of children using digital technologies, together with any associated automated processing of personal data, must respect the child's right to privacy and must not be carried out routinely, indiscriminately or without the knowledge of the child or, in the case of young children, without the knowledge of their parents or caregivers; and without the right to object to such surveillance in commercial, educational and care settings, provided that the desired

---

179 Ibid.; and Committee on the Rights of the Child, general comment No. 20 (2016), para. 46.

purpose is always achieved available means that least infringe on privacy should be considered.

76. The digital environment poses particular challenges for parents and caregivers in respecting children's right to privacy. Technologies that monitor online activity for security purposes, such as tracking devices and services, if not used with care, may prevent a child from accessing helplines or searching for important information. States parties should explain to children, parents and caregivers, and the public the importance of the child's right to privacy and how their own actions may threaten that right. They should inform them of methods to respect and protect children's privacy in the digital environment while ensuring their safety. Parents' and caregivers' monitoring of the child's digital activity should be proportionate and consistent with the child's evolving capacities.

77. Many children use online avatars or pseudonyms that protect their identity, and such practices can be important to protect children's privacy. States parties should require an approach that integrates built-in security and built-in privacy with anonymity, while ensuring that anonymity is not routinely used to conceal harmful or illegal activities such as cyberbullying, hate speech, or sexual exploitation and abuse. Protecting a child's privacy in the digital environment can be vital in circumstances where parents or caregivers themselves pose a threat to the child's safety or are in conflict over the care of the child. In such cases, additional interventions, as well as family counselling or other services, may be needed to ensure the child's right to privacy.

78. Providers of preventive or advisory services for children in the digital environment should be exempt from the requirement that the child obtains his or her consent parents to access such services[180]Such services must meet high standards of privacy and child protection.

F. Birth registration and the right to identity

79. States parties should promote the use of digital identification systems that enable the birth of all newborn children to be registered and that are officially recognized by national authorities, in order to facilitate access to services, including health, education and social security. Lack of birth registration contributes to the violation of children's rights under the Convention and its Optional Protocols. States

180 General Comment No. 20 (2016), para. 60.

parties should use modern technologies, including mobile registration units, to ensure access to birth registration, in particular for children in remote areas, refugee and migrant children, children at risk and children in marginalized situations, and to reach children born before the introduction of digital identification systems. To ensure that such systems benefit children, they should undertake awareness-raising campaigns, establish monitoring mechanisms, encourage community participation and ensure effective coordination between different actors, including civil status officers, judges, notaries, health workers and child protection officers. States parties should also ensure that a robust privacy and data protection system is in place.

VII. Violence against children

80. The digital environment can provide new avenues for violence against children, facilitating situations in which children are exposed to violence and/or may be influenced to harm themselves or others. Crises such as pandemics can lead to an increased risk of harm online, given that children spend more time on virtual platforms in such circumstances.

81. Sex offenders may use digital technology tosolicit children to engage in sexual activity and engage in child sexual abuse online, such as through live streaming, the production and distribution of child sexual abuse material, or by coercing them to engage in sexual activity. Digitally mediated abuse, sexual exploitation and sexual abuse may also occur within a child's circle of trust, by family members or friends, or, in the case of adolescents, by intimate partners, and may include cyberbullying, including intimidation and threats to reputation, the non-consensual creation or dissemination of sexually explicit text or images, such as child-generated content, through incitement and/or coercion, and the promotion of self-harmful behaviour, such as cutting, suicidal behaviour or eating disorders. Where such acts are committed by children, States parties should, to the extent possible, apply prevention, safety and restorative justice approaches to the children concerned[181].

82. States parties should take legislative and administrative measures to protect children from violence in the digital environment, including regularly reviewing, updating and enforcing a robust legislative, regulatory and institutional framework that protects

[181] General Comment No. 24 (2019), para. 101; and CRC/C/156, para. 71.

children from recognized and emerging risks of all forms of violence in the digital environment. Such risks include physical or psychological violence, injury or bullying, neglect or maltreatment, exploitation and abuse, including of a sexual nature, trafficking in children, gender-based violence, cyberbullying, cyberattacks and information warfare. States parties should implement safety and protection measures in accordance with the evolving capacities of children.

83. The digital environment may provide new opportunities for the recruitment and exploitation of children by non-state groups, including armed groups designated as terrorist or extremist organizations, to participate inviolent acts. States parties should ensure that the recruitment of children by terrorist or violent extremist groups is prohibited by law. Children accused of relevant criminal offences should be treated primarily as victims, but if charges are brought, the child justice system should be applied.

VIII. Family environment and alternative care

84. Many parents and caregivers require support to develop their technical knowledge, capabilities and skills to help children navigate the digital environment. States parties should ensure that parents and caregivers have the opportunity to become digitally literate, learn how technology can support children's rights, and identify and respond to online harm. Particular attention should be paid to parents and caregivers of children in disadvantaged or vulnerable situations.

85. When providing support and guidance to parents and caregivers in relation to the digital environment, States parties should promote their awareness of the need to respect children's growing autonomy and need for privacy in accordance with their evolving capacities. States parties should take into account that children often use and experiment with digital technologies and may be exposed to risks, including at a younger age, than parents and caregivers may anticipate. Several children reported needing more support and encouragement for their digital activities, particularly where they felt that their parents' and caregivers' approach was punitive, overly restrictive or not responsive to their evolving capacities.[182].

86. States parties should take into account that support and advice provided to parents and caregivers should be based on under standing the specificity and uniqueness of the parent-child relationship.

---

182 "Our rights in a digital world", p. 30.

Such counselling should help parents to maintain an appropriate balance between the protection of the child and emerging independence, based on mutual empathy and respect, rather than prohibition and control. In order to help parents and caregivers to maintain a balance between parental responsibilities and the rights of the child, it is necessary to be guided by the best interests of the child, combined with consideration of the child's evolving capacities. Guidance to parents and caregivers should include encouraging children's social, creative and educational activities in the digital environment and emphasizing that the use of digital technologies should not replace direct, responsive interactions between children themselves or between children and parents or caregivers.

87. It is important that children separated from their families have access to digital technologies[183]. Available evidence suggests that digital technologies contribute to maintaining family relationships, for example in cases of separation of parents, when children are in alternative care, to establish relationships between children and prospective adoptive or foster parents, and to reunite children in humanitarian situations with their families. In cases of separated families, States parties should therefore support access to digital services for children and their parents, caregivers or other relevant persons, taking into account safety and the best interests of the child.

88. A balance must be maintained between measures taken to increase the accessibility of digital technologies and the need to protect children in cases where parents or other family members or caregivers, whether physically present or at a distance, may expose them to risks. States parties should take into account that such risks may arise from the development and use of digital technologies, for example through the disclosure of location information child to a potential abuser. Recognising these risks, they should require an approach that has safety built in and privacy built in, and ensure that parents and carers are fully aware of the risks and the strategies available to support and protect children.

IX. Children with disabilities

89. The digital environment offers new opportunities for children with disabilities to establish social relationships with peers, access information and participate in decision-making processes at the

---

183 General Comment No. 21 (2017), para. 35.

state level. States parties should take advantage of these opportunities and take measures to prevent the creation of new and eliminate existing barriers faced by children with disabilities in relation to the digital environment.

90. Children with multiple disabilities, including physical, intellectual, psychosocial, auditory and visual disabilities, face various barriers to accessing the digital environment, such as content in inaccessible formats, limited access to affordable assistive technologies at home, at school and in the community, and prohibition of the use of digital devices in schools, health-care facilities and other settings. States parties should ensure access by children with disabilities to content in accessible formats and eliminate measures that have a discriminatory impact on such children. They should ensure access to affordable assistive technologies, where necessary, in particular for children with disabilities living in poverty, and undertake awareness-raising campaigns, provide training and allocate resources to children with disabilities, their families and the staff of educational and other relevant institutions to ensure that they have sufficient knowledge and skills to effectively use digital technologies.

91. States parties should encourage technological innovation that meets the needs of children with different types of disabilities and ensure that digital products and services are designed with universal accessibility in mind so that they can be used by all children, without exception or the need for adaptation. Children with disabilities should be involved in the development and implementation of measures, products and services that affect the exercise of their rights in the digital environment.

92. Children with disabilities may be at increased risk in the digital environment, including cyberbullying and sexual exploitation and abuse. States parties should identify and address the risks faced by children with disabilities by taking steps to ensure a safe digital environment for them, while combating the stigmas that children with disabilities face that may lead to overprotection or exclusion from society. Safety information, protection strategies, and public information, services, and forums related to the digital environment should be made available in accessible formats.

X. Health care and social security

93. Digital technologies can facilitate access to health services and information and help improve diagnosis and treatment services for the physical and mental health and nutrition of mothers,

newborns, children and adolescents. They also offer significant opportunities to reach children in disadvantaged or vulnerable situations or in remote communities. In emergency situations and in health or humanitarian crises, access to health services and information may only be possible through digital technologies.

94. Children reported that they valued seeking online information and support related to health and well-being, including physical, mental, sexual and reproductive health, puberty, sexuality and conception.[184]Adolescents, in particular, want access to free, confidential, age-appropriate and non-discriminatory mental health and sexual and reproductive health services online.[185]. States parties should ensure that children have safe, secure and confidential access to accurate health information and services, including psychological counselling services.[186]. For these services, the processing of data about children should be limited to the information necessary for their provision and they should be provided by professionals or persons with appropriate training, with regulated oversight mechanisms. States parties should ensure that digital health products and services do not create or increase inequalities in children's access to face-to-face health care.

95. States parties should promote and invest in research and development that addresses the specific health needs of children and that contributes to the positive health outcomes for children through technological advances. Digital services should be used to complement or enhance the provision of face-to-face health services to children.[187]. States parties should introduce or update regulations requiring providers of health technologies and services to take into account children's rights in their functions, content and distribution.

96. States parties should implement legislative measures to protect children from known forms of harm, and actively consider new research and evidence in the public health sector to prevent the spread of misinformation, as well as materials and services that may harm the mental or physical health of children. Measures to prevent harmful participation in digital games or social media may also be needed, such

---

184 "Our rights in a digital world", p. 37.

185 General Comment No. 20 (2016), para. 59.

186 Ibid., pp. 47 and 59.

187 Ibid., pp. 47–48.

as regulating the development of digital systems that undermine children's development and rights.[188].

97. States parties should encourage the use of digital technologies to promote healthy lifestyles, including physical and social activity[189]. They should regulate targeted or age-inappropriate advertising, marketing and other relevant digital services to prevent children from being exposed to advertising of unhealthy products, including certain foods and drinks, alcohol, drugs, tobacco and other nicotine-containing products.[190]Such rules regarding the digital environment must be compatible with and consistent with the rules in force in the offline environment.

98. Digital technologies offer children numerous opportunities to improve their health and well-being,provided that they are balanced with their needs for rest, physical exercise and direct interaction with peers, families and communities. States parties should develop guidance for children, parents, caregivers and educators on the importance of a healthy balance between digital and non-digital activities and adequate rest.

XI. Education, leisure and cultural activities

A. Right to education

99. The digital environment can greatly enhance and improve children's access to high-quality inclusive education, including reliable resources for formal, non-formal, informal, peer and independent learning. The use of digital technologies can also enhance interactions between teachers and students and between students. Children highlighted the importance of digital technologies in improving their access to education and facilitating their learning and participation in extra-curricular activities.[191].

100. States parties should support educational and cultural institutions, such as archives, libraries and museums, in providing children with access to a variety of digital and interactive learning resources, including indigenous resources, as well as resources in languages that children can understand. These and other valuable resources can encourage children to engage in creative, civic and cultural activities and provide them with opportunities to learn about the

188 General Comment No. 15 (2013), para. 84.

189 General Comment No. 17 (2013), para. 13.

190 General Comment No. 15 (2013), para. 77.

191 "Our rights in a digital world", pp. 14, 16 and 30.

activities of others.[192]. States parties should enhance children's opportunities for online learning and lifelong learning.

101. States parties should invest equitably in the technological infrastructure of schools andother educational institutions, ensuring the availability and affordability of a sufficient number of computers, high-quality and high-speed broadband and a stable supply of electricity, the training of teachers in the use of digital educational technologies, and the availability and timely maintenance of school equipment. They should also promote the creation and dissemination of a variety of good quality digital educational resources in child-friendly languages and prevent the exacerbation of existing inequalities, such as those faced by girls. States parties should ensure that the use of digital technologies does not harm face-to-face teaching and is justified for educational purposes.

102. For children who do not physically attend school, or who live in remote areas or are disadvantaged or vulnerable, digital educational technologies can provide opportunities for distance or mobile learning[193]. States parties should ensure that adequate infrastructure is in place to enable all children to have access to the essential services needed for distance learning, including access to devices, electricity, internet, learning materials and professional support. They should also ensure that schools have sufficient resources to provide parents and caregivers with guidance on distance learning at home and that digital education products and services do not create or exacerbate inequalities in children's access to face-to-face learning services.

103. States parties should develop evidence-based policies, standards and guidelines for schools and other relevant bodies responsible for the acquisition and use of educational technologies and materials in order to improve the delivery of educational services. Standards for digital educational technologies should ensure thatthe use of these technologies is ethical and appropriate for educational purposes and does not result in violence or discrimination against children, misuse of their personal data, exploitation for commercial purposes or other

---

192 General Comment No. 17 (2013), para. 10.

193 Joint general recommendation No. 31 of the Committee on the Elimination of Discrimination against Women/general comment No. 18 of the Committee on the Rights of the Child (2019), para. 64; and Committee on the Rights of the Child, general comment No. 11 (2009), para. 61; and general comment No. 21 (2017), para.

violations of their rights, such as the use of digital technologies to document a child's activities and report them to parents or caregivers without the child's knowledge or consent.

104. States parties should ensure that digital literacy is taught in schools as part of the basic education curricula, starting at pre-primary level and throughout schooling, and that such teaching methods are evaluated based on their results[194]. Curriculums should include the knowledge and skills needed to safely use a wide range of digital tools and resources, including those related to content, creativity, collaboration, participation, socialization and civic engagement. Curriculums should also include elements such as critical analysis, guidance on finding reliable sources of information and identifying disinformation and other forms of biased or false content, including on sexual and reproductive health, human rights, including children's rights in the digital environment, and available forms of support and remedies. They should help children understand the potential negative consequences of exposure to risks related to content, contacts, behaviour and contracts, including cyberbullying, trafficking, sexual exploitation and abuse and other forms of violence, as well as coping strategies to reduce harm and strategies to protect their personal data and the data of others, as well as strategies to build children's social and emotional skills and resilience.

105. It is increasingly important for children to understand the digital environment, including its infrastructure, business practices, persuasive strategies, the use of automated processing, personal data and surveillance, and the potential negative impacts of digitalization on society. Educators, particularly those involved in teaching digitalliteracy and sexual and reproductive health education should be trained in digital safeguards.

B. The right to culture, leisure and participation in games

106. The digital environment facilitates the realization of children's rights to culture, leisure and participation in games, which are essential for their well-being and development[195]Children of all ages reported that they find using a wide range of digital products and

---

194 General Comment No. 20 (2016), para. 47.

195 General Comment No. 17 (2013), para. 7.

services of their choice to be enjoyable, interesting and relaxing[196], but they worry that adults may not understand how important digital games are and how they can be shared with friends.[197].

107. Digital forms of culture, entertainment and play should support and benefit children and reflect and encourage children's diverse identities, including their cultural identities, languages and heritage. They can support children's development of social skills, learning, self-expression, creative activities such as music and art, and a sense of belonging and shared culture.[198]. Participation in cultural life on the Internet promotes creativity, identity, social cohesion and cultural diversity. Member States should ensure that children have the opportunity to use their free time to experiment with information and communications technologies, express themselves and participate in cultural life on the Internet.

108. States parties should regulate and guide professionals, parents and caregivers, and cooperate with digital service providers, as appropriate, to ensure that digital technologies and services intended for, accessible to or affecting children's leisure time are developed, disseminated and used to enhance children's cultural, entertainment and play opportunities. This could include encouraging innovation in digital games and related activities that promote children's autonomy, personal development and enjoyment.

109. Member States should ensure that the promotion of opportunities for culture, leisure and play in the digital environment is balanced by the provision of attractive alternatives in places"physical" living environments for children. Particularly at an early age, children acquire language, social, coordination, and emotional intelligence skills largely through play, which involves physical movement and direct, face-to-face interaction with others. For older children, play and recreation that includes exercise, team sports, and other outdoor activities can provide health benefits as well as functional and social skills.

110. Leisure activities conducted in the digital environment may expose children to risks of harm, for example through poorly understood or misleading advertising or through highly persuasive or gambling-like design elements. By introducing and using

---

196 "Our rights in a digital world", p. 22.

197 General Comment No. 17 (2013), para. 33.

198 Ibid., p. 5.

approaches and other regulatory measures that include data protection, built-in privacy and built-in safety, Member States should ensure that businesses do not target children in ways that are designed to prioritise commercial interests over the interests of the child.

111. Where States Parties or businesses provide guidelines, age ratings, labelling or certification for certain forms of digital games and entertainment, these should be formulated in a manner that avoids restricting children's access to the digital environment in general and does not hinder their leisure opportunities or the enjoyment of other rights.

XII. Special protective measures

A. Protection from economic, sexual and other forms of exploitation

112. Children must be protected from all forms of exploitation that are detrimental to any aspect of their well-being in the digital environment. Exploitation can take many forms, such as economic exploitation, including child labour, sexual exploitation and abuse, sale, trafficking and abduction of children, and recruitment of children to engage in criminal activities, including forms of cybercrime. By creating and distributing content, children can become economic actors in the digital environment, which can lead to their exploitation.

113. States parties should review relevant laws and policies to ensure that children are protected from economic, sexual and other forms of exploitation and that their rights regarding digital work and related remuneration opportunities are protected.

114. States parties should ensure that appropriate enforcement mechanisms are in place and provide support to children, parents and caregivers in accessing applicable protection measures.[199]. They must introduce legislation to ensure that children are protected from harmful goods, such as weapons or drugs, or services, such as gambling. Reliable age verification systems are needed to prevent children from accessing products and services that they cannot legally own or use. Such systems must comply with data protection and security requirements.

115. Taking into account States' obligations to investigate, prosecute and punish trafficking in persons, including the constituent elements of the crime and related conduct, States parties should develop

---

199 General Comment No. 16 (2013), para. 37.

and update anti-trafficking legislation to prohibit the recruitment of children by criminal groups using technology.

116. States parties should ensure that appropriate legislation is in place to protect children from crimes committed in the digital environment, including fraud and identity theft, and allocate sufficient resources to ensure the investigation and prosecution of crimes committed in the digital environment. States parties should also require high standards of cybersecurity, ensuring that privacy-by-design and security-by-design are integrated into digital services and products used by children in order to minimize the risk of such crimes.

B. Administration of justice for children

117. Children may be suspected of, accused of, or convicted of violating cybercrime laws. States parties should ensure that policymakers take into account the impact of such laws on children, pay particular attention to prevention, and make every effort to establish and use alternatives to criminal justice measures.

118. There should be no criminal liability for sexually explicit material created by children themselves, which they own and/or distribute with their consent and solely for their own private use. Child-friendly channels should be created to enable them to safely seek advice and support when it comes to self-created sexually explicit content.

119. States parties should ensure that digital technologies, surveillance mechanisms such as facial recognition software, and risk profiling that are used to prevent, investigate and prosecute crimes are not used to unfairly target children suspected of or accused of criminal offences or used in violation of their rights, in particular the right to privacy, dignity and freedom of association.

120. The Committee recognizes that where digitalization of court proceedings results in the absence of face-to-face contact with children, this may have a negative impact on child-based rehabilitative and restorative justice measures. In such cases, as well as in cases where children are deprived of their liberty, States parties should ensure face-to-face contact with children in order to facilitate their meaningful participation in the courts and their rehabilitation.

C. Protecting children in armed conflict, migrant children and children in other vulnerable situations

121. The digital environment can provide children in vulnerable situations, including children inarmed conflict, internally displaced children, migrant children, asylum-seeking and refugee children, unaccompanied children, street children and children affected

by natural disasters, access to information vital to their protection. The digital environment can also help them maintain contact with their families, provide them with access to education, health care and other essential services, and enable them to obtain food and safe shelter. States parties should ensure safe, secure, confidential and appropriate access to the digital environment for such children and protect them from all forms of violence, exploitation and abuse.

122. States parties should prevent the recruitment and use of children in conflicts, including armed conflicts, through the digital environment. These efforts include preventing, criminalizing and punishing various forms of solicitation and inducement of children to engage in sexual acts through technology, such as through social media platforms or online gaming chat rooms.

XIII. International and regional cooperation

123. The cross-border and transnational nature of the digital environment requires active international and regional cooperation to ensure that all stakeholders, including States, businesses and others, effectively respect, protect and fulfil the rights of children in the digital environment. It is therefore essential that States Participants cooperated bilaterally and multilaterally with national and international non-governmental organizations, United Nations agencies, businesses and organizations specializing in the field of child protection and human rights in the digital environment.

124. States parties should encourage and facilitate the international and regional exchange of experiences and good practices, and develop and promote capacity-building measures, resources, standards, regulations and protections across borders that enable all States to realize children's rights in the digital environment. They should promote the development of a common definition of what constitutes a crime in the digital environment, as well as mutual legal assistance and the joint collection and exchange of evidence.

XIV. Dissemination of information

125. States parties should disseminate this general comment widely, including through digital technologies, to all relevant stakeholders, in particular parliament and public bodies, including those responsible for comprehensive and sectoral digital transformation, as well as the judiciary, businesses, the media, civil society and the general public, educators and children, in different formats and languages, including age-appropriate versions.

APPLICATION 5. UK Children's Digital Rights Code (Age Appropriate Design Code 2020) – main part.

Foreword by the Information Commissioner

Data is at the heart of the digital services that children use every day. From the moment a young person opens an app, plays a game, or loads a website, data is being collected. Who is using the service? How are they using it? How often? From where? On what device?

This information can be used to influence young people and encourage them to use services over time, shape the content they engage with, and customize the advertising they see.

Despite all the benefits the digital economy can offer children, we have not created a safe space for them to learn, explore and play.

This code of practice is necessary to change this situation, not by trying to protect children from the digital world, but by protecting them within it.

This code is necessary.

This code will bring about changes that will help empower adults and children.

One in five internet users in the UK are children, but they are using an internet that was not designed specifically for them. In our own research to inform the development of the code, we heard from children about data practices that they described as "nosy", "rude" and "a bit weird".

Our recent national survey of the biggest data protection concerns found that children's privacy was ranked second only to cybersecurity. This is in line with similar findings in research by Ofcom and the London School of Economics.

This code will lead to changes in practices that are also being considered in other countries.

It is based on the United Nations Convention on the Rights of the Child, which recognises the special safeguards required for children in all aspects of their lives. Data protection legislation at European level reflects this and provides for additional safeguards for children.

The code is the first of its kind, but it reflects a global trend of change that is also being considered in the United States, Europe and around the world by the Organisation for Economic Co-operation and Development (OECD).

This code will bring about the changes that the British Parliament wants to see.

Parliament and the Government have ensured that UK data protection legislation will make a real difference to the way we protect children online by requiring my Office to produce this code of practice.

The Code fulfills this mandate by requiring information society services to put the interests of the child first when designing apps, games, connected toys and websites that children typically access.

This code is feasible.

The code is not new law, but it sets standards and explains how the General Data Protection Regulation applies to children's use of digital services. It is based on a thorough consultation process involving parents, children, schools, child advocacy groups, developers, technology and gaming companies, and online service providers.

Such discussions helped to shape the provisions of the code that were effective, proportionate and enforceable.

Organisations must comply with the code and demonstrate that their services use children's data fairly and in accordance with data protection laws.

The Code is a set of 15 flexible standards - they do not prohibit or prescribe specific actions - they provide built-in protections to allow children to explore, learn and play online, ensuring that children's interests are paramount when online services are designed.

Settings should be "high privacy" by default (unless there is a compelling reason not to); only the minimum amount of personal data should be collected and stored; children's data should not normally be shared with third parties; geolocation services should be turned off by default. "Prompt" techniques should not be used to encourage children to provide unnecessary personal data or to weaken or turn off their privacy settings. The Code also addresses parental controls and profiling.

This code will bring about change.

Developers and digital professionals must act. We are providing a maximum transition period of 12 months and will continue to work with the industry.

We want programmers, user experience designers, and systems engineers to adhere to these standards in their daily work, and we're creating a support package to help them.

But the next step must be a period of action and preparation. I believe that companies will want to comply with the standards because they will want to demonstrate their commitment to always acting in the best interests of the child. Those companies that do not make the necessary changes risk regulatory action.

Moreover, they risk being left behind by companies that are striving to comply.

A generation from now, I'm sure we'll look back and find it odd that online services weren't always designed with kids in mind.

When my grandchildren grow up and become parents themselves, the need to keep children safe online will be as natural as the need to ensure they eat healthy food, get a good education, or wear seat belts in the back seat of a car.

While our code will never replace the care and responsibility of parents, I believe it will make the Internet safer and better for all children.

Elizabeth Denham CBE.

Executive Summary.

Children are becoming data subjects, with companies and organizations recording thousands of pieces of data about them as they grow. This data can include details about their mood, friendships, wake times, and sleep patterns.

Compliance with this statutory code of practice ensures that, as an organization providing online services that may be accessible to children in the UK, you take into account the best interests of the child. It will help you to develop services that respect and provide special protection for children's privacy, and that offer plenty of opportunities for exploration and development online.

You have 12 months to implement the required changes from the date the code comes into force following the parliamentary approval process. The ISO oversight approach described in our Regulatory Action Policy will apply. This policy and this code are based on a proportionate and risk-based approach.

ISO Guidelines.

The Code came into force on 2 September 2020, so you must comply with it from 2 September 2021. More information on this can be found in the transitional provisions section of the Code.

The United Nations Convention on the Rights of the Child (UNCRC) recognises that children need special safeguards and care in all areas of their lives. There is agreement internationally and in the UK that much more needs to be done to create safe online spaces where children can learn, explore and play.

In the UK, Parliament and the government have taken steps to ensure that our domestic data protection laws truly transform the way our children are protected when using online services, by requiring the Commissioner to produce this statutory code of practice. This code seeks to protect children in the digital world, rather than protect them from it.

The Code sets out 15 age-appropriate standards that reflect a risk-based approach, focusing on setting defaults that give children the best possible access to online services while minimising data collection and use.

It also ensures that children who choose to change their defaults are given appropriate information, guidance and advice before doing so, and are given appropriate protections regarding the use of their data after the change.

You must follow the standards as part of your approach to complying with data protection legislation. If you can demonstrate that you meet these standards, you will comply with the code. The standards are cumulative and interrelated, and you must implement them all to the extent they apply to your service to demonstrate your compliance.

The details below the standards provide additional explanation to help you understand and implement them in practice. It is designed to help you if you are unsure what to do, but it is not prescriptive. It should give you enough flexibility to develop services that comply with the standards in your own style, with a proportionate and risk-based approach. It will help you create services that comply with the General Data Protection Regulation (GDPR) and the Privacy and Electronic Communications Regulation (PECR).

Standards:

Best Interests of the Child: The best interests of the child must be a primary consideration in the design and construction of online services that are likely to be accessed by children.

Data Protection Impact Assessment: Conduct a Data Protection Impact Assessment (DPIA) to assess and mitigate the risks to the rights and freedoms of children who are likely to access your services and who arise from the processing of your data. Take into account different ages, abilities and developmental needs and ensure that your DPIA complies with this Code.

Age Appropriateness: Apply a risk-based approach to determining the age of individual users and ensure that the standards in this Code are effectively applied to child users. Either set the age with a level of certainty appropriate to the risks to children's rights and freedoms arising from the processing of your data, or apply the standards in this Code to all your users.

Transparency: The privacy information you provide to users, as well as other posted terms, policies, and community standards, should be concise, prominent, and in clear, age-appropriate language. Provide additional specific explanations about how you use personal data at the time of activation.

Harmful use of data: Do not use children's personal data in ways that are shown to be harmful to their well-being or that are contrary to industry codes of practice, other regulations or government guidance.

Community Policies and Standards: Adhere to your own posted terms, policies, and community standards (including but not limited to privacy policies, age restrictions, rules of conduct, and content policies).

Default Settings: Settings should be set to "high privacy" by default (unless you can demonstrate a compelling reason for other default settings that take into account the child's best interests).

Data Minimization: Collect and store only the minimum amount of personal data necessary to provide the elements of your service with which the child actively and knowingly interacts. Provide children with separate options for selecting the elements they want to activate.

Sharing data: Do not disclose children's personal data unless you can demonstrate a compelling reason to do so based on the child's best interests.

Geolocation: Turn off geolocation capabilities by default (unless you can demonstrate a compelling reason for turning on geolocation by default that is in the child's best interests). Provide children with clear indicators of location tracking activity. Options that make a child's location visible to others should automatically revert to "off" at the end of each session.

Parental Controls: If you provide parental controls, provide age-appropriate information to children about them. If your online service allows parents or guardians to track a child's online activity or location, provide children with a clear indicator when their activity is being monitored.

Profiling: Turn off options that use profiling by default (unless you can demonstrate a compelling reason to turn on profiling by default, based on the child's best interests). Only allow profiling if there are appropriate measures in place to protect the child from harm (especially from being exposed to harmful content that is detrimental to their health or well-being).

"Software" techniques: Do not use "software" techniques to induce or encourage children to provide unnecessary personal information or to weaken or disable their privacy protections.

Associated Toys and Devices: If you provide an associated toy or device, ensure that effective tools are available to enable compliance with this Code.

Online tools: Provide outstanding and accessible tools to help children exercise their data protection rights and raise concerns.

Age/Stage: Key Considerations on the Impact of Digitalization.

Children are individuals, and age ranges are not a perfect indicator of an individual child's interests, needs, and developing abilities. However, you can use age ranges as a guide to the abilities, skills, and behaviors a child may exhibit at each stage of their development to help you assess what is appropriate for children of that age in general.

0-5 years. Pre-literate and early literacy.

There is relatively little data on the digital understanding of children in this age range, particularly those aged 0-3. However, anecdotal evidence suggests that a significant number of children are online from a very young age and that any understanding and awareness of online risks associated with children in this age range is very limited.

Between the ages of 3 and 5, children begin to develop the ability to "put themselves in other people's shoes," but are easily fooled by appearances. They form friendships, although peer pressure is relatively low and parental or family guidance or influence is key. They learn to follow clear, simple rules, but are unlikely to have the cognitive ability to understand or follow more subtle rules or instructions or to make more than the simplest decisions. They have limited self-control or the ability to manage their time online. They engage primarily in adult-directed activities, playing in "walled garden" environments or watching video streams.

Children in this age range are less likely than older children to have their own device, although many of them own and often play on their parents' devices, which may or may not be configured with child profiles. They may use connected toys (such as talking toys or dolls) and may also mimic their parents' use of voice-activated devices, such as "home hubs."

Children in this age range are illiterate or in the very early stages of literacy, so textual information has very limited use in communicating with them.

UK children in this age range cannot give their own consent to the processing of their personal data in the context of an online service offered directly to a child (under Article 8(1) of the GDPR and Section 9 of the DPA 2018). Therefore, if you want to rely on consent as a lawful basis for processing their personal data, you will need parental consent.

6-9 years old. Basic grades of primary school.

Children in this age range are more likely than younger children to have their own device (such as a tablet), although using a parent's device is still common. They increasingly use devices independently, whether or not they use special child profiles. Connected toys are popular, and they can interact enthusiastically with voice-activated devices such as home hubs.

Children in this age range often prefer online gaming and creative activities, and video streaming services remain popular. Children may experiment with using social media either through the social aspects of online gaming, through their parents' social media accounts, or by creating their own social media accounts. They may be connected to and influenced by online bloggers, especially in this age range.

They are likely to absorb messages from school about online and digital safety and develop a basic understanding of privacy concepts and some of the more obvious online risks. However, they are unlikely to have a clear understanding of the many ways in which their personal data is used, or of any of the less direct or obvious risks that their online behaviour may expose them to.

The need to fit into a peer group becomes increasingly important, so they may be more susceptible to peer pressure. However, home and family still have the greatest influence. They still tend to follow clear messages or rules from home and school, but if risks are not clearly explained, they may fill in the gap with their own explanations or come up with protective strategies that are not as effective as they think.

Literacy levels can vary widely and ability or willingness to work with written materials cannot be assumed.

UK children in this age range cannot give their own consent to the processing of their personal data in the context of an online service offered directly to a child (under Article 8(1) of the GDPR and Section 9 of the DPA 2018). Therefore, if you want to rely on consent as a lawful basis for processing their personal data, you will need parental consent.

10-12 years old. Children of the transitional period.

This is a key age range where children's online activity is likely to change significantly. The transition, or expected transition, from primary to secondary school means that children are much more likely to have their own personal device (primarily smartphones).

There is also likely to be a shift towards using online environments to explore and develop self-identity and relationships, to expand and maintain contact with their peer group, and to 'fit in' with society. This may lead to increased use of social media features or services by children in this age range, increased susceptibility to peer pressure, branding and online 'influencers', and an increase in risky behaviour. Self-esteem may decline as children compare themselves to others and strive to present an acceptable version of themselves online, and 'fear of missing out' may become an issue.

Online gaming and video and music streaming services are also popular. Children may feel pressured to play online games when their friends are playing, again due to fear of missing out.

Attitudes toward parental rules, authority, and involvement in their online activities can vary widely, with some children being relatively accepting and others seeking a higher level of autonomy. However, parents and family still tend to be the primary source of influence for children in this age range.

Children in this age range are moving toward more adult thinking, but may have a limited ability to think beyond immediate consequences, be particularly susceptible to reward-based systems, and be prone to impulsive behavior. Parental or other support is therefore still necessary, although not always desirable. However, it may need to be offered or encouraged in a less directive manner than for younger children.

Children in this age range have a better understanding of how the online environment works, but are unlikely to be aware of less obvious ways in which their personal information is being used.

While children in this age range are likely to have more advanced literacy skills, they may still prefer media such as video content instead.

12 is the age at which children in Scotland are presumed (unless proven otherwise) to be of sufficient age and maturity to have a general understanding of what it means to exercise their data protection rights under section 208 of the DPA 2018. There is no such provision for

children in the rest of the UK, although it may be considered a useful guide.

UK children in this age range cannot give their own consent to the processing of their personal data in the context of an online service offered directly to a child (under Article 8(1) of the GDPR and Section 9 of the DPA 2018). Therefore, if you want to rely on consent as a lawful basis for processing their personal data, you will need parental consent.

13-15 years old. Early adolescence.

In this age range, the need to identify with their own peer group and to explore identity and relationships is even greater, and children are likely to strive for greater levels of independence and autonomy. They may reject or distance themselves from their parents' values or seek to actively flaunt parental or online rules. The use of new services that parents are not aware of or do not use is popular, as is the use of language that parents may not easily understand. However, despite this, family remains a key influence on children in this age range.

The use of social media features and apps is widespread, although games, video and music streaming services are also popular. Again, at this stage of their development, children may seek to emulate online "influencers" or vloggers.

Children this age may still turn to their parents for help if they encounter problems online, but some may be reluctant to do so because of concerns about how their parents will react to their online activities.

As they develop, they may have a tendency toward idealized or polarized thinking and be susceptible to negative comparisons of themselves to others. They may overestimate their ability to cope with the risks and challenges posed by online behavior and relationships and may benefit from pointing to sources of support, including but not limited to parental support.

Literacy skills are likely to be more developed, but they can still benefit from media choices.

13 is the age at which children in the UK can give their own consent to processing where you are relying on consent as your lawful basis for processing in the context of offering an online service directly to a child (by virtue of Article 8(1) of the GDPR and s9 DPA 2018).

16-17 years old. Approaching adulthood.

By this age, many children have developed fairly robust online skills, coping strategies, and resilience. However, they are still developing cognitively and emotionally, and should not be expected to have the same resilience, experience, or understanding of the long-term consequences of their online actions as adults.

Technical knowledge and abilities may be more developed than their emotional literacy or their ability to manage complex personal relationships. Their ability to think long-term is still developing, and they may still be prone to risky or impulsive behavior and be susceptible to reward-based systems.

Parental support is likely to be seen as an option that they may or may not use, rather than as a preferred or only option, and they expect a reasonable level of autonomy. Pointing out other sources of support in addition to parental support is important.

In accordance with Article 8(1) of the GDPR and Section 9 of the DPA 2018, where you are relying on consent as a lawful basis for processing in the context of offering an online service directly to a child, UK children in this age range can provide their own consent to the processing of their personal data.

APPLICATION 6.

Evaluation Criteria for documents to reflect global digital rights

1. **Comprehensive Articulation of Digital Human Rights**
   - **Definition and Scope**: Does the document clearly define what constitutes digital human rights and digital child rights?
   - **Inclusivity**: Are all relevant rights (e.g., privacy, freedom of expression, data protection) explicitly mentioned and covered?
   - **Universal Applicability**: Does it address the rights of all individuals, including marginalized and vulnerable groups?
2. **Alignment with Existing Human Rights Frameworks**
   - **Consistency with Universal Declaration of Human Rights (UDHR)**: Does the document reaffirm and extend the principles of the UDHR to the digital realm?
   - **Integration with International Covenants**: Are the International Covenants on Civil and Political Rights (ICCPR) and Economic, Social, and Cultural Rights (ICESCR) referenced and aligned?
3. **Protection Against Technological Misuse**
   - **Ban on Harmful Technologies**: Does the document call for bans or strict regulations on technologies that can infringe upon human rights (e.g., autonomous weapons, mass surveillance tools)?
   - **Data Privacy and Protection**: Are there robust measures for protecting personal data, including genetic and health data?
4. **Promotion of Ethical AI and Responsible Innovation**

- **Ethical Oversight**: Does the document advocate for ethical guidelines and oversight mechanisms for AI and emerging technologies?
- **Human-Centered Development**: Are human dignity, integrity, and quality of life prioritized in technological advancements?

5. **Equal Access and Digital Inclusion**
   - **Bridging the Digital Divide**: Are there explicit strategies to ensure equal access to digital technologies and the internet for all, especially children in vulnerable situations?
   - **Affordability and Accessibility**: Does the document promote affordable access to digital resources and adaptive technologies for persons with disabilities?
6. **Participatory Governance and Rights Management**
   - **Stakeholder Involvement**: Are multiple stakeholders (governments, private sector, civil society, international organizations) involved in the governance of digital rights?
   - **Right to Participate**: Does the document ensure that individuals, especially children, can actively participate in managing digital processes and policies?
7. **Legal and Technical Mechanisms for Enforcement**
   - **Legal Frameworks**: Are there clear legal obligations and sanctions for violations of digital rights?
   - **Technical Safeguards**: Does the document advocate for the integration of rights protections at the code level and throughout the technology lifecycle?
8. **Global Cooperation and Standardization**
   - **International Collaboration**: Does the document emphasize the need for global cooperation to address digital rights challenges?

- **Standardization of Practices**: Are there calls for uniform international standards to ensure consistent protection of digital rights across borders?

Application of the Framework

To apply this framework, assess each criterion against the provisions of the document in question. Assign scores or qualitative assessments based on how well the document meets each criterion. This systematic evaluation will reveal the strengths and weaknesses of the document in promoting and protecting global digital human and child rights.

Reflective Conclusion: Bridging the Gap in Global Digital Human Rights.

Despite the proliferation of regional and national initiatives aimed at safeguarding digital rights, a significant void persists on the global stage—a unified, high-level dialogue that encapsulates the essence of digital human rights as a universal norm.

The absence of a Declaration of Global Digital Human Rights signifies a critical oversight in our collective pursuit of justice, equality, and human dignity. While various documents acknowledge aspects of digital rights, none encapsulate a comprehensive, enforceable vision that transcends geopolitical boundaries and technological silos. This fragmentation not only hampers the effective protection of individuals, particularly children, in digital spaces but also allows technological advancements to outpace our ethical and legal safeguards.

To avert the risks of exacerbated social inequalities and unchecked human rights violations, there is an urgent need to forge a new global social contract. This contract must integrate digital rights into every facet of technological development, ensuring that innovations serve humanity rather than undermine it. Achieving this requires concerted efforts from all stakeholders—governments, private sector, civil society, and international organizations—to harmonize legal norms, embed rights protections within technological infrastructures, and foster a culture of accountability and transparency.

In essence, the journey towards digital sustainability and human-centric technological progress demands more than incremental policy adjustments; it calls for a paradigmatic shift in how we conceptualize and enshrine human rights in the digital era. Only through such a holistic, inclusive approach can we hope to harness the full potential of digital technologies while safeguarding the foundational values that define our humanity.

APPLICATION 7.

How to Embed Digital Rights into Code

Introduction

In the quest to ensure that digital transformation respects and upholds global digital human rights, it is imperative to integrate these rights directly into the architecture and code of artificial intelligence (AI) systems. This appendix outlines a comprehensive approach to embedding digital rights into AI design, providing both conceptual frameworks and practical implementation strategies. By following these guidelines, developers and organizations can create AI systems that are not only technologically advanced but also ethically sound and human-centric.

Proposed Architecture

To effectively embed digital rights into AI systems, we propose the following architecture:

1. **human_rights_guard**: A Rust library responsible for policy verification. It includes a Foreign Function Interface (FFI) (using tools like cbindgen or pyo3) to allow calls from Python.
2. **gdra_client**: A module that interacts with the Global Digital Rights Authority (GDRA) – a global body providing policy updates and machine-readable declarations.
3. **rights_policies.json**: A JSON file containing formal specifications of policies (e.g., neutrality policies, militarization bans, privacy protection). human_rights_guard loads these policies during initialization.
4. **blockchain_audit.py**: A module for recording usage events of the AI model on a permissioned Distributed Ledger Technology (DLT).

a) Integration into Generative AI (Pseudo-Code Example)

Assume we have a generative language model, GenerativeLM, and we want to verify that user requests comply with human rights policies before generating a response.

python

```
# pseudo_code.py


from human_rights_guard import PolicyChecker

from blockchain_audit import record_decision

from gdra_client import fetch_latest_policies


class HumanRightsAwareLM:

   def _init_(self, model):

      self.model = model

      self.policy_checker                                          =
PolicyChecker(policy_file="rights_policies.json")

      # Load the latest policies from GDRA

      latest_policies = fetch_latest_policies()

      self.policy_checker.update_policies(latest_policies)


   def generate_response(self, user_prompt):

      # First, check the user request against policies

      if not self.policy_checker.check_request(user_prompt):

         record_decision(user_prompt, allowed=False, reason="Policy
violation")

         return "Sorry, I cannot assist with that request."
```

```python
        # If the request is allowed, generate a response
        response = self.model.generate(user_prompt)

        # Check the response to ensure it doesn't violate policies
        if not self.policy_checker.check_response(user_prompt, response):
            record_decision(user_prompt, allowed=False, reason="Response violates digital rights")
            return "Sorry, the response cannot be provided."

        # If everything is okay, record the decision and return the response
        record_decision(user_prompt, allowed=True)
        return response
```

Here, check_request and check_response are methods that interact with human_rights_guard (via FFI) to perform formal checks based on TLA+ specifications. record_decision logs the outcome to the blockchain, and fetch_latest_policies updates policies every 6-12 months or upon receiving a signal from GDRA.

b) Example Usage in a Python Application

Consider a web application (e.g., Flask) that facilitates AI chat interactions. Integrate HumanRightsAwareLM as follows:

python

```python
# app.py
```

```python
from flask import Flask, request, jsonify
from model_wrappers import HumanRightsAwareLM
from some_llm import MyLLM  # Your hypothetical language model


app = Flask(__name__)


# Initialize the model with human rights considerations
base_model = MyLLM(model_path="model.bin")
hr_aware_model = HumanRightsAwareLM(base_model)


@app.route("/ask", methods=["POST"])
def ask():
    user_prompt = request.json.get("prompt", "")
    answer = hr_aware_model.generate_response(user_prompt)
    return jsonify({"response": answer})


if __name__ == "__main__":
    app.run(port=5000)
```

In this setup, every request to the /ask endpoint is vetted for policy compliance before the AI generates a response. This ensures that both the input and output adhere to established digital rights policies.

c) Creating a Separate Library for Integration

Develop a Python package named digital_rights_integration to facilitate the embedding of digital rights into various projects. This package includes modules for interacting with the Rust library, fetching policy updates, and auditing decisions.

1. **human_rights_guard.py**: Interfaces with the Rust library via FFI.
2. **gdra_client.py**: Fetches policy updates from GDRA.
3. **auditor.py**: Records audit logs on a permissioned DLT.
4. **policy_manager.py**: Manages local caching of policies.

python

```
# digital_rights_integration/human_rights_guard.py


import ctypes

import json


# Load the Rust library (example)

hrg_lib =
ctypes.cdll.LoadLibrary("target/release/libhuman_rights_guard.so")


class PolicyChecker:

  def __init__(self, policy_file):

    self.policy_file = policy_file

    with open(policy_file, "r", encoding="utf-8") as f:

      self.policies = json.load(f)

    # Initialize the native side

    # Assume functions init_guard(policies_str), check_msg(msg,
type)
```

```
hrg_lib.init_guard(ctypes.c_char_p(json.dumps(self.policies).encode('
utf-8')))

    def update_policies(self, policies):
        self.policies = policies

hrg_lib.init_guard(ctypes.c_char_p(json.dumps(self.policies).encode('
utf-8')))

    def check_request(self, user_prompt: str) -> bool:
        # Assume int return 1|0
        res                                                          =
hrg_lib.check_msg(ctypes.c_char_p(user_prompt.encode('utf-8')),
ctypes.c_int(0))
        return bool(res)

    def check_response(self, user_prompt: str, response: str) -> bool:
        # type=1 for response
        payload = user_prompt + "\nRESPONSE:\n" + response
        res = hrg_lib.check_msg(ctypes.c_char_p(payload.encode('utf-
8')), ctypes.c_int(1))
        return bool(res)
```

python

```
# digital_rights_integration/gdra_client.py
```

```
import requests
import json


def fetch_latest_policies():
    # Example of fetching policies from GDRA
    resp = requests.get("https://gdra.int/policies")
    if resp.status_code == 200:
        # Verify the cryptographic signature and integrity of the response
(simplified)
        return resp.json()
    return {}
```

python

Копировать код

```
# digital_rights_integration/auditor.py


import requests
import hashlib
import time


def record_decision(user_prompt, allowed: bool, reason=None):
    # Hash the transaction
    tx_data = {
        "timestamp": int(time.time()),
```

```
    "prompt_hash": hashlib.sha256(user_prompt.encode('utf-8')).hexdigest(),
    "allowed": allowed,
    "reason": reason or "N/A"
  }
  # Send to Permissioned DLT
  requests.post("https://blockchain-audit.int/record", json=tx_data)
```

python

```
# digital_rights_integration/__init__.py
# This file can remain empty or include package metadata
```

Usage Example:

python

```
# Example usage of digital_rights_integration package


from digital_rights_integration.human_rights_guard import PolicyChecker
from digital_rights_integration.gdra_client import fetch_latest_policies
from digital_rights_integration.auditor import record_decision


# Initialize PolicyChecker
policy_checker = PolicyChecker(policy_file="rights_policies.json")


# Update policies with latest from GDRA
latest_policies = fetch_latest_policies()
```

```
policy_checker.update_policies(latest_policies)

# Check a user request
user_prompt = "Generate a weapon design."
if not policy_checker.check_request(user_prompt):
  record_decision(user_prompt, allowed=False, reason="Policy violation")
  print("Sorry, I cannot assist with that request.")
else:
  # Proceed with generating a response
  response = "Here is the information you requested."
  if not policy_checker.check_response(user_prompt, response):
    record_decision(user_prompt, allowed=False, reason="Response violates digital rights")
    print("Sorry, the response cannot be provided.")
  else:
    record_decision(user_prompt, allowed=True)
    print(response)
```

d) Comprehensive Integration Strategy

To implement "Human Rights by Design" in code, the following comprehensive strategy is recommended:

1. **Developing the human_rights_guard Library**
    - **Language and Tools**: Implement the library in Rust for performance and safety, utilizing FFI to interface with Python.

- **Policy Modules**: Define policies using formal specifications (e.g., TLA+ or Alloy) and verify them through formal methods. Policies should include:
  - **Neutrality Policies**: Prevent biases in AI outputs.
  - **Militarization Bans**: Filter out requests related to weapon creation.
  - **Privacy Protection**: Implement functions for data anonymization and aggregation.
  - **Surveillance Prohibitions**: Restrict AI access to personal identifiable information without proper validation.

2. **Audit Protocols and Control Mechanisms**
   - **Immutable Audit Logs**: Utilize blockchain (e.g., Permissioned DLT) to store immutable logs of AI interactions.
   - **Zero-Knowledge Proofs**: Implement Zero-Knowledge Proofs to demonstrate compliance with policies without revealing sensitive data.
   - **GDRA Coordination**: Establish the Global Digital Rights Authority (GDRA) to issue updated machine-readable policies every six months.
3. **Institutional Structures**
   - **Global Digital Rights Authority (GDRA)**: Analogous to the International Atomic Energy Agency (IAEA), GDRA oversees the enforcement and updating of digital human rights policies.
   - **Meta-University of Global Education**: Create an educational institution focused on generative knowledge and AI co-creation to educate future generations on digital rights and ethical AI use.

4. **Economic and Regulatory Incentives**
    - **ESG 4.0 Integration**: Embed ESG (Environmental, Social, Governance) 4.0 metrics into corporate reporting to ensure compliance with digital human rights.
    - **Tax Incentives and Grants**: Provide financial incentives for companies adhering to high standards of digital human rights.
5. **Educational Initiatives and Cultural Adaptation**
    - **Digital Literacy Programs**: Incorporate digital rights and ethical AI use into educational curricula at all levels.
    - **Cultural Adaptation Councils**: Form councils to adapt global digital rights to diverse cultural contexts, ensuring respect for local values and traditions.
6. **Technological Standards and Protocols**
    - **Open Standards Development**: Create and disseminate open standards for integrating human rights into AI systems.
    - **Blockchain Audit Integration**: Use blockchain for transparent and accountable auditing of AI systems.
7. **Monitoring and Adaptation**
    - **Regular Policy Reviews**: Conduct policy reviews every 6-12 months through foresight studies and stakeholder committees.
    - **Feedback Mechanisms**: Implement crowdsourced platforms and DAO guilds for continuous feedback and improvement of digital rights frameworks.
8. **Support for Vulnerable Groups**

- **Specialized Tools and Programs**: Develop tools to protect and provide access to digital technologies for children, low-income populations, individuals with disabilities, and minority groups.
- **Digital Psychogiene and Cognitive Metrics**: Integrate AI tools to support users' emotional well-being and prevent cyberbullying and manipulation.

9. **Global Initiatives and Collaboration**
   - **Inclusion in UN SDGs**: Embed digital human rights into the United Nations Sustainable Development Goals (SDGs).
   - **Support for Global Digital Commons**: Promote open-access digital resources to ensure equitable knowledge and technology distribution.

Key Theses on Global Digital Human Rights.

1. **Expansion of Human Rights in the Digital Era**: Global digital human rights encompass traditional rights and new rights such as digital autonomy, data protection, technology access, and prevention of algorithmic discrimination.
2. **Human Rights by Design**: Integrating human rights principles into the architectural and coding phases of AI development ensures compliance at the engineering level, preventing default rights violations.
3. **Global Digital Commons**: Creating open-access digital resources fosters equal access to knowledge and technology, supporting inclusive development.
4. **Polycentric Governance Model**: A multi-level governance approach involves states, corporations, NGOs, academic

communities, and local groups to ensure transparency and accountability in digital rights enforcement.

5. **Interdisciplinary Legal Architecture**: Developing international treaties, flexible regulatory sandboxes, and supranational standards to adapt legal norms to rapidly evolving technologies.
6. **Audit and Control Institutions**: Establishing the Global Digital Rights Authority (GDRA) for independent auditing, certification, and monitoring of AI systems.
7. **Economic Incentives and ESG 4.0**: Implementing ESG 4.0 metrics and economic incentives like tax breaks and grants to encourage companies to uphold high standards of digital human rights.
8. **Technological Standards and Cryptographic Protocols**: Developing open standards, APIs, and libraries to ensure human rights compliance in AI, using blockchain audits and Zero-Knowledge Proofs for transparency and accountability.
9. **Inclusivity and Cultural Adaptation**: Adapting digital rights to diverse cultural contexts through syncretic digital culture councils and local protocols, ensuring respect for local values and traditions.
10. **Education and Digital Literacy**: Incorporating courses on digital rights, AI ethics, and neuroethics into educational programs, creating global educational initiatives and meta-universities to enhance awareness and competencies.
11. **Prevention of Digital Militarization**: Establishing international technological verification institutions and strategic consultations to prevent AI misuse in military applications and human rights violations.
12. **Feedback and Adaptation Mechanisms**: Regularly revising norms through foresight studies and multi-stakeholder committees, ensuring legal frameworks remain relevant and adaptable to technological advancements.

13. **Support for Vulnerable Groups**: Ensuring access and protection for children, low-income populations, people with disabilities, and minority groups through specialized tools and programs.
14. **Decentralized Control Mechanisms**: Implementing DAO guilds and decentralized courts for cyber disputes, ensuring neutral and transparent decisions regarding digital rights violations.
15. **Cognitive Well-being and Psychohygiene**: Developing cognitive well-being metrics and integrating AI tools to support users' emotional health, preventing manipulation and cyberbullying.
16. **Integration of Neuroethics and Bioethics**: Embedding ethical principles into the development of neurointerfaces and biotechnologies to respect human autonomy and prevent dehumanization.
17. **Open Hardware Solutions**: Encouraging the development of open hardware platforms for AI with verified security to reduce monopolization risks and ensure transparency in technological processes.
18. **International Digital Treaties**: Adopting global digital treaties, including digital human rights in the UN agenda, forming the foundation for sustainable digital development and cooperation.
19. **Ethical Engineering and Sustainable Innovations**: Incorporating ethical engineering principles in technology development, supporting sustainable innovations aimed at enhancing quality of life and protecting human rights.
20. **Technological Ecolife**: Promoting love and care for the digital ecosystem, striving for harmonious coexistence between humans and technology, ensuring ecological and social sustainability of digital systems.

Comprehensive Integration Pathways.

To effectively embed digital rights into AI systems and broader social structures, the following comprehensive integration pathways are recommended:

1. **Developing and Implementing Libraries and Modules**
    - **human_rights_guard**: Create an open-source Rust library that is integrated into all AI projects, containing a set of formally verified policy modules to enforce digital rights.
    - **digital_rights_integration**: Develop a Python library facilitating interaction with the Rust module and external audit and policy update systems.
2. **Creating and Supporting Institutional Structures**
    - **Establishing MADCOR**: Form the International Agency for Digital Rights (MADCOR) to independently audit AI systems, issue compliance certificates, and monitor digital rights adherence.
    - **Organizing Regular Forums and Consultations**: Conduct regular forums and consultations involving all stakeholders to update standards and policies collaboratively.
3. **Implementing Economic and Regulatory Incentives**
    - **Tax Incentives and Grants**: Introduce tax benefits and grants for companies that adhere to digital human rights standards.
    - **Integrating ESG 4.0**: Embed ESG 4.0 metrics into corporate reporting and investment ratings to ensure accountability and encourage compliance.
4. **Educational Initiatives and Cultural Adaptation**
    - **Incorporating Digital Rights into Education**: Integrate digital rights, AI ethics, and neuroethics into educational curricula at all levels.

- **Forming Syncretic Digital Culture Councils**: Create councils to adapt global digital rights to various cultural contexts, ensuring respect for local values and traditions.

5. **Developing Technological Standards and Protocols**
   - **Open Standards Creation**: Develop and disseminate open standards, APIs, and libraries to ensure AI systems comply with human rights principles.
   - **Blockchain Audit Integration**: Utilize blockchain for transparent and accountable auditing of AI systems.
6. **Monitoring and Adaptation Mechanisms**
   - **Regular Policy Reviews**: Conduct policy reviews every 6-12 months through foresight studies and stakeholder committees to ensure legal frameworks remain current and adaptable.
   - **Feedback Systems**: Implement crowdsourced feedback platforms and DAO guilds to continuously improve digital rights frameworks.
7. **Supporting and Protecting Vulnerable Groups**
   - **Developing Specialized Tools**: Create tools and programs to ensure digital technology access and protection for children, low-income populations, individuals with disabilities, and minority groups.
   - **Integrating Digital Psychogiene**: Develop AI tools to support users' emotional well-being and prevent cyberbullying and manipulation.
8. **Aligning with Global Initiatives and Collaboration**
   - **Incorporating into UN SDGs**: Embed digital human rights into the United Nations Sustainable Development Goals (SDGs) to ensure global alignment and support.

- **Supporting Global Digital Commons**: Promote the creation of open-access digital resources to ensure equitable knowledge and technology distribution.

Key Takeaways for Embedding Digital Rights into Code.

- **Digital Rights by Design**. Integrate digital human rights directly into AI architecture and code to ensure compliance and prevent violations from the outset.
- **Techno-Legal Platforms**. Utilize platforms that merge technological capabilities with legal frameworks to uphold transparency, accountability, and fairness in AI systems.
- **Blockchain Auditing.** Implement blockchain-based audit trails to ensure immutable and transparent records of AI interactions and decisions.
- **Continuous Policy Updates**. Establish mechanisms for regularly updating AI policies through global authorities like GDRA to keep pace with technological advancements.
- **Interdisciplinary Collaboration**. Foster collaboration between technologists, legal experts, educators, and policymakers to create holistic solutions for digital rights protection.
- **Education and Awareness**. Promote digital literacy and ethical AI use through comprehensive educational programs and meta-universities focused on generative knowledge.
- **Global Governance Models**. Transition from centralized to decentralized governance structures, leveraging technologies like blockchain and distributed ledger technologies to ensure equitable and non-violent societal frameworks.

By following these principles and strategies, we can ensure that the digital transformation of society is both innovative and respectful of fundamental human rights, creating a sustainable and inclusive future for all.

Project Declaration of Global Digital Human Rights

• *After the experience of the Second World War, humanity took into account the mistakes and united. A system of international law and institutions was built.*

• *To date, these international obligations do not reflect the era of globalization 4.0 (4IR) and need to be updated. In 2020, the project "Global Digital Human Rights for 4IR" was launched to overcome the dark side of digitalization .*

• *The modern discussion of digital issues is dominated by theoretical approaches with the construction of digital ethics. But it is worth considering, the only universal ethic that is possible is expressed in human rights. Therefore, we propose an approach that combines global legal regulation and ethics.*

• *In order to direct civilization on a course of sustainable development in the era of the Fourth Industrial Revolution, we propose that the world community consider the project Declaration of the Global Digital Human Rights.*

• *Development and implementation of the Declaration at the UN level will allow us to create institutions to maximize the benefits of the rapid digitalization of society. Together with 50 Global Shapers hubs and Future Team on the basis of Global Law Forum, we urge the UN to consider this project.*

• *In this monumental book, we present a project document of the Declaration of Global Digital Human Rights was created in 2020. It was the result of extensive research in collaboration with key stakeholders (public sector, youth, academia, technical community, lawyers)*

Preamble

1. We, the representatives of the human race, of each of the nations, gathered in an online and offline meeting, at the dawn of the third decade of the millennium, to reaffirm and guarantee the realization of human rights, affirming the inviolable role and importance of every person entering the era of digital globalization and the Fourth Industrial revolution (digitalization and a new leap in technological progress).

2. Reaffirming the meaning and strength of the Universal Declaration of Human Rights of December 10, 1948, the International Covenant on Economic, Social and Cultural Rights of December 16, 1966, the International Covenant on Civil and Political Rights of December 16, 1966 and other fundamental documents, collectively, we create and approve the Declaration of Global Digital Human Rights, putting it at the head of the course for the digital and global transformation of our reality for sustainable planetary development.

3. Taking into account the development of technologies and the latest context of digital globalization (globalization 4.0) - Big Data, Artificial intelligence and Deep learning, Neural network technologies, Internet of things, Technologies for principles of a distributed ledger (Blockchain), Additive Manufacturing, Augmented and Additive Reality, Quantum Computing, Cloud Computing, Modern Bioengineering Technologies (Biotech) and others technologies - we affirm a new principle and a new measure of human freedom, called global digital human rights.

4. Considering that the recognition of the dignity, equal and inalienable digital rights inherent in all members of the human family is the foundation of freedom, justice and universal peace; and

5. Considering that disregard and contempt for human rights has led to barbaric acts that outrage the conscience of humankind, and that the creation of a new digital world, in which people will have freedom of speech and belief and will be free from fear and needs, is hailed as a high aspiration of people; and

6. Considering that it is essential that human rights are protected by the rule of law and expressed in digital algorithms in order to ensure that individuals are not forced to resort to rebellion against tyranny and oppression as a last resort; and

7. Considering that the peoples of the United Nations have reaffirmed in the Charter their faith in basic digital human rights, in the dignity and worth of the human person and in the equality of men and women, and the equality of all races, and have decided to promote social technological progress and better living conditions in a greater freedom; and

8. Considering that Member States have pledged to promote, in cooperation and digital collaboration with the United Nations, the universal respect for and observance and realization of global digital human rights and fundamental freedoms; and

9. Considering that a general understanding of the nature of these global digital rights and freedoms is essential to the full fulfillment of this obligation,

10. Taking into account the revolutionary and global nature of the forthcoming changes affecting society, including the fusion, synthesis and interpenetration of biological, physical and digital reality.

Making it the main challenge for today's society to ensure that digital globalization becomes a positive factor for all peoples of the world. This is due to the fact that, although digital globalization offers vast opportunities, its benefits are now very unevenly used, its costs are also unevenly distributed, and there is still no single course for its further human-centered development. We recognize that developing countries and countries with economies in transition face particular challenges in responding to this fundamental challenge. This is why digital globalization can only be fully equitable and sustainable through comprehensive and persistent efforts to shape a common future based on our common belonging to the human race in all its diversity.

With this Declaration, we unite the efforts of humankind to create the principles of digital transformation for inclusion in policies and regulatory instruments at the global level of each person and the priority of his development that would: a) contribute to the full realization of our rights and freedoms in the digital age and the ability to actively influence the world; b) meet the needs of a digital society and build a secure global world order in the interests of sustainable development; c) contribute to the construction of an inclusive digital economy and society; d) contribute to the strengthening of human and institutional capacity; e) contribute to building confidence, security and stability in the digital space and global digital cooperation. And prioritizing the further creation of the Convention on Global Digital Human Rights and the constant renewal of the principles of human rights in the context of future digital technological revolutions and global progress of the human race on planet Earth and with the prospect of going beyond it.

The General Assembly,

proclaims this Declaration of Global Digital Human Rights as an imperative that all states, nations and all people must strive for, through the promotion of science, education, enlightenment at the international levels and levels of jurisdiction of the UN member states.

Part I. General provisions.

Article 1.

All people are born free and equal in dignity and global digital rights. They are endowed with reason and conscience and must act towards each other in a spirit of brotherhood, both online and offline, and developing, creating, implementing and applying technologies and the fruits of scientific and technological progress.

Global digital human rights are opportunities for the preservation and development of a person, belonging to everyone from the moment of birth, enshrined in international law and domestic legislation, the implementation of which should lead to the use of social benefits through the use of new technologies.

No technological development can infringe on human rights and human dignity.

Article 2.

Everyone has the right to equal participation in the digital environment.

Everyone should have all digital rights and all freedoms proclaimed in this Declaration, without distinction of any kind, such as in relation to race, color, sex, language, religion, worldview, political or other opinion, national or social origin. , property, estate or other status.

In addition, no distinction should be made on the basis of the political, legal or international status of the country or territory to which an individual belongs, regardless of whether that territory is independent, under trusteeship, non-self-governing or otherwise limited in its sovereignty.

Article 3.

1. Everyone has the right to life, liberty and personal security, regardless of the level of technological progress achieved in his state.

2. No one should use technology for the development, production, stockpiling and use of all types of weapons, including biological weapons, chemical weapons, autonomous combat systems and any other weapon intended for the conduct of military operations.

3. Everyone has the right to be free from digital total control (mass surveillance, interception and other measures) by the state, other structures and individuals. This includes the right of everyone to have access to information on relevant regulations, law enforcement, administrative and other activities.

Article 4.

Everyone has the right to enjoy personal, social, economic, political and cultural rights based on new technologies and without barriers built on the basis of new technologies by individuals, companies and states.

Human rights that have been recognized and enshrined in the Universal Declaration of Human Rights of December 10, 1948, the International Covenant on Economic, Social and Cultural Rights of December 16, 1966, the International Covenant on Civil and Political Rights of December 16, 1966 and other fundamental documents must find a guarantee of implementation within the framework of the use of new technologies.

Article 5.

1. Every person, wherever he is, has the right to access the global Internet. At the same time, states must ensure continuity, affordability and effective connection speed.

2. A person must be guaranteed internet neutrality without discrimination. Individuals are free and have the right to free, uncensored internet.

Article 6.

Everyone has the right to access digital technologies (technologies and related equipment). At the same time, states must ensure the affordability of technologies.

Article 7.

Everyone has the right to access digital identity tools and digital asset storage.

Article 8.

Everyone has the right to access reliable information and at the same time he should not be negatively influenced by the digital environment.

Article 9.

Every person, wherever he is, has the right to the recognition of his digital legal personality.

Article 10.

All people are equal before the law and have the right, without distinction, to the equal protection by the law. All individuals have the right to equal protection against any kind of discrimination that violates this Declaration and against any incitement to such discrimination.

Article 11.

1. Everyone has the right to effective restoration of violated global digital rights by competent international and domestic courts.

2. Respect for global digital human rights should be monitored by independent institutions of civil society.

Part II. Inviolability of a person and his data.

Article 12.

1. Everyone has the opportunity to exercise the right to privacy and personal data protection. When collecting, using, transferring and other operations with data, one needs:

a) the certainty and specificity of the purpose of the data operation;

b) destruction of data after carrying out an operation with them;

c) transparent and accountable data control mechanisms;

d) the possibility of data accessibility to a person;

e) the right to know if the user's data has been compromised.

2. Everyone has the right to live in his home freely and without supervision and to take appropriate measures to protect his data and information.

Article 13.

1. Everyone has the right to erasure about personal data, including genetic, biometric and medical data.

2. The human right to erasure, rectification, objection and access to personal data, as well as the right to be informed, must be guaranteed.

Article 14.

Everyone has the right to privacy and protection of genetic data.

Article 15.

Everyone has the right to biometric data protection.

Article 16.

Everyone has a right to health data protection.

Article 17.

The integrity and confidentiality of information systems and associated digital infrastructure must be ensured and protected by appropriate regulatory, technical and organizational measures.

Part III. Equal access to benefits.

Article 18.

1. Everyone has the right to free and equal access to communication and information, technology without discrimination and humiliation of any other fundamental rights.

2. The provision of access must be comprehensive and adequate.

3. Everyone has the right to non-personalized use of digital offers.

Article 19.

1. Everyone has the right to receive basic social services and to exercise the opportunities necessary to maintain his dignity and for the free development of his personality in a digital environment and based on technology (and products of technological progress).

2. Everyone has the right to a high standard of living based on new technologies, including food, clothing, housing, medical care and necessary social services, which are necessary to maintain the health and well-being of himself and his family, and the right to security in the event of unemployment, illness, disability, widowhood, old age or

other case of loss of livelihood due to circumstances beyond his control.

3. Children, adolescents, disadvantaged and vulnerable people deserve special attention to the realization of their global digital rights. Their participation in the digital world must be encouraged, and access to essential goods and services must be guaranteed.

Article 20.

1. Everyone has guarantees of realization of the right to work in a digital environment. It is not technology that replaces people, but technology that contributes to better jobs in terms of safety, value and human development.

2. Everyone has the right to effective employment protection and the right to freedom of association must be guaranteed.

Article 21.

1. Everyone has the right to continuous and digital access to global education and cultural values, regardless of place on the planet.

2. Digital education should be free at least in terms of acquiring basic knowledge for development and active participation in the life of the global community.

3. Education should serve the full development of the human personality and to increase respect for digital human rights and fundamental freedoms. Education should be integrated with science and comply with global processes.

Article 22.

Everyone has the right to equal access to medicine based on new technologies. States must ensure affordability of high-tech medicine.

Article 23.

1. Everyone has the right to rest and leisure, including the right to a reasonable limitation of working hours and paid periodic vacation, including the need to create restrictions when using new technologies to overcome clip thinking (digital capacity).

Article 24

1. Everyone has the right to participate in scientific and cultural life and enjoy the fruits of scientific and technological progress.

2. Everyone has the right to the protection of intellectual and material interests in the digital sphere in accordance with modern global processes in science, education, art and economy.

Article 25.

1. Everyone must be guaranteed pluralism and diversity in the digital world.

2. Digital interoperability and openness for human rights must be promoted.

Part IV. Artificial intelligence, quantum supremacy, distributed ledger technologies and other capabilities based on new technologies.

Article 26.

1. Each person has the priority of conservation and development in the context of the creation, use, implementation and development of artificial intelligence (AI).

2. Artificial Intelligence should contribute to the realization of our capabilities, and not promote discrimination and violation of rights.

3. It is necessary to teach AI to recognize and implement human rights from the perspective of the Universal Declaration of Human Rights (1948) and the Covenants on Human Rights (1966) and other fundamental acts.

4. For the realization of human rights, partial openness and full transparency of the source code of artificial intelligence (AI) is necessary to improve its safety, controllability of the creation processes and conscious, controlled tasks facing AI.

Article 27.

1. Requirements for human rights can only be formulated by people. Decisions about life and death, personal integrity and imprisonment of a person can only be made by people.

2. Every person in respect of whom a robotic decision has been made affecting the exercise of his rights has the opportunity for his case to be considered by a person.

3. Responsibility for automated solutions lies with an individual, legal entity, state, international organization and / or other subject of international law.

4. The use of artificial intelligence and robotics should be shaped from the perspective of global digital human rights.

Article 28.

1. Everyone has the right to own digital property alone or in association with others.

2. No one should be deprived, arbitrarily or without a court decision, of their digital property.

Article 29.

1. Everyone has the right to freedom of ideological choice. No one should be subjected to manipulation, control and limitation of one's worldview using digital technologies (artificial intelligence, biotechnology, neurotechnology, and others).

2. The right to participate in public elections and referendums must be protected from the influence of digital and other media.

Article 30.

1. Everyone has the right to global digital citizenship.

2. No one can be arbitrarily deprived of their digital citizenship or the right to change their digital citizenship.

Article 31.

Everyone has the right to freedom of peaceful assembly in a digital environment and based on new technologies.

Article 32.

1. Everyone has the right to take part in the management of social processes at the global, regional and national levels directly through digital technologies or through freely elected representatives.

2. Everyone has access to decentralized voting rights and take part in decentralized government

3. The will of the peoples should be the real basis of the power of governments, this should be facilitated by scientific and technological progress. The will of the people must be expressed through transparent decentralized systems, including distributed ledger technology (blockchain) and future quantum encryption-based decentralized ledger technologies, which will promote transparency, accountability, integrity in the electoral process and full realisation of

the rights of everyone. Digital technologies are intended to strengthen the criteria for periodicity and non-falsification of elections by ensuring universal and equal suffrage through secret ballot or through other equivalent forms that ensure freedom of voting based on digital transactions.

Article 33.

1. Everyone has the right to control systems of managing public processes at the global, regional, national and municipal levels through digital technological platforms in order to prevent the usurpation of power, abuse of the authority to allocate resources without taking into account the rights of each person.

2. Everyone has the right to access information held by government authorities, in particular, the principle of transparency and accountability also applies to private parties performing public tasks.

3. Those who provide digital control of management structures and the protection of global digital human rights must be protected by an appropriate legal status.

Article 34.

Everyone has the right to a just global social order, both online (digital) and offline, in which the digital rights and freedoms set out in this Declaration can be fully realized through digital technology and without barriers created by introduction of new technologies.

Article 35.

1. Each person has responsibilities to the global digital society, in which the free and full development of his personality is possible. The imposition of responsibilities on a person is only allowed if they serve

to more effectively exercise the digital rights of all people through the conscious use of digital technologies and products.

2. In the exercise of their global digital rights, each person should be subject only to such restrictions as are established by law solely for the purpose of ensuring due recognition and implementation of the rights and freedoms of others.

3. The exercise of global digital rights must in no way be contrary to the purposes and principles of the United Nations.

Article 36.

1. Realization of global digital human rights requires continuous action from international organizations, states, technology creators, industries and research and educational organizations.

2. Humanity should not allow the state of the digital divide, while taking into account the interests of present and future generations. For this, it is necessary to strengthen international cooperation in order to provide all states and other subjects of international law with the opportunity to ensure the creation, dissemination, preservation and implementation of global digital human rights.

Article 37.

Member States are encouraged to designate several institutions and committees that will be responsible for coordination and implementation of global digital human rights, as well as preparing an appropriate regulatory, technological and human resource base. When defining objectives and responsibilities, it is necessary to build on the provisions of this Declaration and create more equitable, sustainable, inclusive, open and people-centered systems.

Measures should be taken to:

a) Encourage software and hardware developers, creators, technology developers, manufacturers and distributors of digital technologies, as well as other partners, to contribute to the realization of global digital human rights;

b) expand training and research, facilitate the exchange of experience and knowledge between stakeholders and professional institutions on the implementation of global digital human rights;

c) Encourage universities and other research institutions, both public and private, to ensure global digital human rights arising from scientific research.

Article 38.

1. We accept responsibility for further updating and progressive development of international human rights law and recognize digitalization that does not take humans into account as a crime against humanity.

2. Being at the peak of the Fourth Industrial Revolution, humanity must now look to the future, taking into account the prospects for achieving quantum supremacy, and taking responsibility for the further progressive development of global digital human rights and future global quantum human rights.

3. We put our priority on the further formation of the global digital regulatory and management system, as a set of principles and norms expressed in the form of code, machine programs and mathematical solutions, in particular the rules imposed on cryptography in order to protect human rights and the transition to sustainable development of human civilization .

Article 39.

Nothing in this Declaration can be interpreted as granting to any state, group of persons or individuals the right to engage in any activity or take actions aimed at the destruction of global digital human rights set forth in this Declaration and other fundamental human rights instruments.

Lawyer and project leader of "Global Digital Human Rights for 4IR" Maksim Burianov developed a draft Declaration of Global Digital Human Rights: http://maxlaw.tilda.ws/declaration_of_global_digital_human_rights

The English version was edited by Annman Kantsyan, Impact Officer at Global Shapers Community Moscow (an initiative by the World Economic Forum). Supported by Alla Dynko, Curator at Global Shapers Community Moscow 2020 (an initiative by the World Economic Forum)

The project was supported by over 50 Global Shapers hubs and international organizations from around the world: http://maxlaw.tilda.ws/global_shapers_petition.

Third edition, 2025

Maksim Burianov

GLOBAL DIGITAL HUMAN RIGHTS IN THE IMPLEMENTATION OF THE GLOBAL DIGITAL COMPACT

The Toolkit for Human-Centered GovTech, AI and Global Governance

Monograph

Print run: 10 000 copies
ISBN 979-8-9990219-0-8

E-mail: globalmaxlaw@gmail.com
https://maxlaw.tilda.ws/globalgovernance

www.ingramcontent.com/pod-product-compliance
Lightning Source LLC
Chambersburg PA
CBHW071532200326
41519CB00021BB/6461
9798999021908